The Architecture of Bridges

THIS BOOK HAS BEEN

PRODUCED UNDER A GRANT

FROM THE AMERICAN BRIDGE COMPANY,

A SUBSIDIARY OF

THE UNITED STATES STEEL CORPORATION

Elizabeth B. Mock

THE **ARCHITECTURE** OF

Bridges

THE MUSEUM OF MODERN ART · NEW YORK

Reprint Edition 1972   *Published for The Museum of Modern Art by Arno Press*

Library of Congress Catalog Card Number 70-169309
ISBN 0-405-01568-2

CONTENTS

Acknowledgments

This book was started three years ago, when I was Curator of the Museum of Modern Art's Department of Architecture, and was undertaken at the suggestion of Philip L. Goodwin, trustee of the Museum, who for years had been urging the Department to make an attempt to raise the level of American bridge design. What started merely as a gesture of friendship soon became an absorbing interest, so I am grateful to Mr. Goodwin for opening my eyes to bridges. Other sponsors were Philip C. Johnson, now Director of the Museum's Department of Architecture and Design, and Edgar Kaufmann, Jr., Advisor to that Department. To Mr. Johnson I am also greatly indebted for editorial suggestions, particularly as concerned with the special problems of a picturebook. I wish to thank Mary Barnes, former Curator of Architecture, and Ruth L. Bookman, former Assistant Curator, for their assistance in gathering material and their criticism of the text. The photograph reproduced on the paper jacket was suggested by Edward Steichen, Director of the Photography Department.

Outside the Museum there is a long list of people who have been kind and helpful to me in this rather presumptuous project, and to whom I wish to express abiding gratitude: Rudolf Mock, for needed encouragement and advice as well as for his contribution of the diagrams that illustrate *Structural Types*; Henry-Russell Hitchcock, Jr., for provocative conversation on the subject of nineteenth-century iron bridges and the loan of relevant material; Adolf Meyer, Chief of the Tennessee Valley Authority's Civil Design Branch, for invaluable criticism of the text; Sigfried Giedion, for the stimulus of his long and articulate championship of the Swiss engineer, Maillart, also for the use of some photographs which he assembled for the Museum of Modern Art's traveling exhibition of Maillart's work; Paul Zuberbühler, for material on old and new Swiss bridges; Max Bill, for his photographs of Maillart bridges; G. E. Kidder Smith, for his material on Swedish bridges; Bernard Rudofsky, for his critical reading of the introduction; Marcel Fornerod, for information about Freyssinet's prestressed concrete construction.

I am similarly indebted to André Bloc, editor of *L'Architecture d'aujourd'hui*, Alfred Roth, editor of the Swiss magazine, *Das Werk*, Waldo Bowman, editor of the *Engineering News Record*, Elizabeth Fitten, of Princeton University's Marquand Library, and R. E. Enthoven, of the Library of the Royal Institute of British Architects. Other libraries that extended special courtesies were the Technical Library of the Tennessee Valley Authority, the Lawson McGhee Public Library of Knoxville, Tennessee, the Engineering Societies Library of New York, the Architectural Library of Harvard University, and the main library of Princeton University. The picture collection of the American Institute of Steel Construction was extremely helpful in the preliminary research, and I wish particularly to thank the editors of the *Engineering News Record* for the loan of innumerable photographs from their excellent files.

Above all others I am beholden to Frank Lloyd Wright, whose influence has prompted me to search in bridges for the qualities of organic architecture. But the interpretation is entirely my own, and I am alone responsible for its inadequacies.

Publication of the book was made feasible by a generous subsidy from the American Bridge Company, granted upon the recommendation of Mr. J. H. Zorn, of that company, and Mr. Robert J. Ritchey, of the United States Steel Corporation.

E.B.M.

Taliesin West
March, 1949

In an old graveyard of Concord, Massachusetts, is a slab with a date, 1791, and an inscription:

In Memory of Captain John Stone
the Architect of that Modern and
Justly Celebrated Piece of Architecture
Charles River Bridge

A similar conviction that a fine bridge is also fine—and modern—architecture, is the basis of this book and its only justification.

Bridges are architecture, but architecture of a very special kind, unique in its single-mindedness. Ordinarily the art of architectural or landscape design consists in the creation of space, and structure is finally the means to that end. But since the function of a bridge is simply the continuation of a roadway over a void, its structure is both means and end, and its reality lies not in space enclosed, but in structure itself. Since a bridge does not define space, but cuts through it, it is free of all the intricate psychological considerations that must be taken into account when space is molded or enclosed. Thus, paradoxically, a bridge is at once the most tangible and most abstract of architectural problems. As such, it is capable of extraordinary purity, though it may perhaps never achieve the richness and depth of expression that are possible in buildings of more complex human motivation.

Since the reality of a bridge lies in its structure, the art of bridge building lies in the recognition and development of the beauty latent in those structural forms that most effectively exploit the strength and special properties of a given material.

Beauty is not automatic; technical perfection alone is not enough. A great engineer is not a slave to his formulas. He is an artist who uses his calculations as tools to create working shapes as inevitable and harmonious in their appearance as the natural laws behind them. He handles his material with poetic insight, revealing its inmost nature while extracting its ultimate strength through structure appropriate to its unique powers.

Today we boast the most powerful materials of all time: steel and steel-reinforced concrete. But there is a curious reluctance to explore their ultimate possibilities and accept their full esthetic implications—a reluctance based on the idea that massiveness is itself a virtue, as it was in the days when stone was the only strong, permanent, therefore honorable material. Arch-reactionaries in this sense were the Nazis. Needing heavily pretentious buildings to symbolize the immortal glory of their State, they developed this characteristically specious line of reasoning: "Form requires mass; without mass, no artistic, architectural form; without form, no beauty." They cited the ruptured Tacoma span (page 63) as evidence that "Abolishment of mass leads not only to formlessness but to failure." German engineers in actuality paid scant attention to this facile official theorizing, and bridge design was therefore the one field of art in which the Third Reich was not completely tripped up by its own mock-heroics.

The identification of beauty with mass has never been as deliberate as it was in Germany in the thirties, but the two are very often confused. Even when this confusion is unconscious, it is a very real obstacle to the achievement and acceptance of quality in contemporary bridge design. The old stone-builders set themselves no such limitation. On the contrary, they were constantly seeking new ways to lighten their spans by making every stone a lively working element of the structure, and thus to minimize the massiveness of what was, after all, a massive material. Man has rarely built less efficiently than he was able, and the history of bridge architecture is essentially the story of his triumph over space through increasingly skillful exploitation of the best materials available to him. His triumph was not only over space, but over the inertia of material. The nineteenth-century engineers were working within this grand tradition when they welcomed the new possibilities for efficiency—and for delight—that were offered by metal. They made the sudden transition from the massiveness of stone to the finely etched lines of iron and steel with magnificent assurance. Only in the last fifty years or so have bridges been overtaken by nostalgia for the reassuring weightiness of stone construction, and their forms falsified accordingly. The varying relationship between architecture and bridge design has had something to do with this change of heart.

Until comparatively recently bridges were similar in genesis to other types of architecture; that is, the more important were designed by architects, the less pretentious by anonymous local craftsmen. This fruitful unity of the structural arts was broken in the eighteenth century by the invention of engineering as a separate, highly specialized profession. The split between science and art was not abrupt. In fact, they continued on generally friendly, mutually sympathetic terms through the early nineteenth century. Gradually, however, architecture tended to deteriorate into mere decoration, and architects stewed contentedly in their own precious juices of stylistic revival and eclecticism, divorced from the reality of the great new building problems and the great new materials that were to solve them.

It has been customary to lament the break between architecture and engineering, but amateurs of nineteenth-century metal bridges should be grateful that the Battle of the Styles was not fought over a bridge-head. For the engineers, seeking no justification in historical precedent, were free to find appropriate expression for the new materials. And they were free to create new esthetic values through the revealed energy and the almost miraculous lightness of their gravity-defying spans. Thus the best work of the engineers was more truly architecture—in the proper sense of that word—than the nostalgic re-creations and adaptations of those who called themselves architects.

The twentieth century has witnessed a curious reversal of position. Unlike the great engineers of the preceding century, who saw that bridge construction and bridge esthetics were inseparable, and felt quite capable of solving both, together, today's engineer rarely looks beyond his ever more formidable trove of scientific and technical knowledge. Contemptuous of art, he tends to satisfy himself with mere expediency. When specially called upon for beauty, he usually, either by himself or with the advice of a decorator-architect, seeks to embellish his indifferent structure with some kind of external "styling," thus confusing whatever clean, inherently expressive lines his design might originally have possessed. This "styling" assumes many guises, but most often it is an attempt to recall the massiveness of stone construction—sometimes even its specific shapes, especially that of the time-hallowed arch. It is a strange fact that lightness is more readily accepted in horizontals than in verticals. Horizontal members may be distorted into arches, masked with a stone shell, or laden with vulgar ornament, yet they are rarely deliberately thickened. But it is with real gusto that today's run-of-the-mill bridge designer sets about the work of dramatizing the solidity of his vertical supports, particularly if these supports are the abutments of real or apparent arches. It is worth noting that the architects, who invented the ingenious cosmetics now cherished by the engineers, are themselves rapidly learning to discard them. Contemporary architects are assimilating the lesson of nineteenth-century engineering; perhaps contemporary engineers would profit from a study of the principles of twentieth-century architecture.

It would not be fair, however, to blame the faked massiveness of most of today's bridges entirely upon the engineers and their consulting decorators. Often it is popular and official taste that is culpable. Even the great Maillart (pages 102-113), when he was finally allowed to build a bridge in Switzerland's capital city of Berne, was forced to sacrifice his proposal of a lithe and elegant three-hinged arch to the official demand for a massive stone-like vault.

Such superficial beautification is far more common in the United States than in Europe. And what is more, American bridges are actually grossly over-dimensioned as compared with their European counterparts. That is not wholly the fault of the American engineer, for he works under a terrible handicap: *American materials are too cheap.* Europe, with its historic pattern of relatively expensive material and relatively cheap labor, has been pushed into extremely economical design and the inventior of new and ever more efficient ways of building. Scarcity of material has also encouraged good craftsmanship; when one has a single stick of wood one handles it with love and care.

In the United States conditions are quite the opposite. Competent labor comes high; and fine spare design, with its demand for careful computation, special steelwork, highly skilled workmanship and conscientious supervision, is almost prohibitively expensive. It is much cheaper just to throw in a few more yards of concrete, a few more tons of steel. But the advantages of economy of material are proportionate to the length of a span, and enormous spans are feasible only when the "dead load" of the material itself is reduced to a minimum. This was the special incentive for the magnificent slenderness of our great suspension bridges—our one important contribution to modern bridge design.

Almost any American is alert to the airy beauty of these suspension bridges. Curious, then, that he should distrust lightness in other types of bridge design, and that he should look to massiveness for his pleasure. Sometimes he tries to justify his prejudice by claiming that lightly drawn bridges look unrelated to their surroundings. But that argument has small validity. For when the scale of a bridge seems wrong the fault is almost always one of brutality. Massive concrete arches, for example, can dwarf and distort a man-scaled urban or rural scene in most distressing fashion; and even in grandiose natural terrain, where a heavy structure might seem justified, the contrast of a delicately membered bridge may be far more delightful. The spidery trusses of nineteenth-century viaducts are as good a case in point as the tenuous lines of our own suspension bridges. Far better merely to ask that a bridge disturb its surroundings as little as possible than to seek an over-literal harmony.

Economy of material cannot alone assure design excellence. Nor is it enough to add a demand for justice of proportions, for the refinement of structural elements and the clarification of their relationships. A bridge can be much more than the sum of these rather negative virtues. It can be the bringing together of material, structure and form as one thing, one song in space.

Integrity in this sense is inherent, not imposed. It is not a question of paring down, nor has it to do with mere arrangement or composition. It seems to come only from the conception of structure as an organism developing according to the law of its own nature, quickening inert material into life and giving it meaning.

This esthetic ideal is technically substantiated and physically invigorated by the relatively new idea of structural continuity. When structure is continuous, a bridge is no longer an assemblage of separately computed, separately functioning items. Instead, all elements act together, literally fused into a single working shape. It was through continuity of structure that the plastic nature of reinforced concrete was first made explicit, in Maillart's bridges. These prophesy a future in which welded steel and plastic-bonded plywood, like reinforced concrete, will be molded into thin shells, stiffened by bending. Abandoning line for surface, skeleton for shell, right angles for curves, and two dimensions for three, a bridge will become, more than ever before, a single splendid gesture dedicated to the conquest of space.

As the mad Caligula knew when he recklessly launched a bridge into the Mediterranean, a beautiful bridge has a life quite beyond its purely practical functions.

There are only three basic types of bridge construction—beam, arch and suspension cable. Combinations of these are possible, but the bridge that follows one unified, clearly defined principle of construction is generally more satisfactory in appearance than the hybrid. Unfortunately for the amateur observer, however, the structural system of a bridge, even when pure, is not always immediately recognizable, for it depends less on superficial form than on the manner in which the load is transferred from the span to the points of support. In other words, an arch is not necessarily an arch.

Beams neither thrust nor pull: they rest. Their load is transmitted vertically to the supports, and gravity is the only force involved. Easy examples are the stone clapper bridge (page 12), the primitive log bridge (page 30), the early iron trussed girders (page 70), and the reinforced concrete overpass pictured on page 117.

Everyone knows from experience that a simple beam tends to bend and break at the middle of its span. This means that the lower part of a beam is subject to pull or tension, even while the upper portion is being squeezed together or compressed. Beam construction is therefore appropriate only to materials strong in both tension and compression: stone beams, though feasible, make little sense, as stone lacks tensile strength.

In modern multi-span bridges the beams are often allowed to run continuously over a number of supports. These *continuous beams* use their materials so efficiently that they can be relatively shallow and light. And unlike the simple one-span beam, their greatest strength is generally required over the intermediate points of support. Therefore their depth can logically be decreased toward the center of each span, exactly where extra headroom is often advantageous. The examples illustrated on pages 67 to 69 and 118 to 119 show that the lower edge of the beam may be brought down to the support in either a diagonal or a curve; in the latter case the bridge tends to look like a series of arches, though there is no arch action involved.

A variation of the beam principle is found in the *cantilever* bridge. A cantilever is essentially just a beam that projects out beyond its vertical support or supports. Sometimes two symmetrical cantilevered structures are set arm to arm, perhaps forming an arched opening as in the viaduct at Viaur (page 78). More often the cantilever arms do not themselves meet, but are connected by a light "suspended span." The classic example of this highly developed version of cantilever construction is the great Firth of Forth Bridge (page 77). A similar system of cantilevered beam ends and suspended spans is advantageous under certain conditions for continuous beam bridges.

The *rigid frame* (or portal frame, as it is sometimes called) is a special case, for the beam is monolithic with its supports, and horizontals and verticals form a rigid unit. Here again, the depth of the span may profitably be reduced toward the center and the transition effected in a smooth curve. The result may be an arch in appearance, but it is still a beam in action. An example is the Dry Creek Bridge on page 116.

Arches are much more lively than beams, for they are constantly pushing outward against their supports or abutments. Since the load of an arch is transferred diagonally rather than vertically to its supports, the planes of contact must be inclined, and the abutments must be strong enough to meet the powerful thrust of the arch. The arch itself is squeezed or compressed, therefore well adapted to construction in a material strong

only in compression: stone, or unreinforced concrete. Depending upon the firmness of the ground, the arch is either fixed at each end or provided with two or three hinges to allow for possible movement.

The *fixed arch* may be of uniform depth throughout, but in its most expressive form it is shallowest at the crown (the top) and grows deeper as it approaches its abutments. Fine examples are the wooden footbridge on page 38, the Russian Gulch Bridge (page 89) and Freyssinet's famous through-arch at St. Pierre du Vauvray (page 99). Stone arches are invariably fixed.

The *two-hinged arch* is provided with a hinge at each abutment, where its load is concentrated in a point. Its most dramatic form is the crescent or sickle-shaped arch, best illustrated by Eiffel's viaduct at Garabit (page 43).

The *three-hinged arch* allows for movement at the crown as well as at each end. In unskilled hands it tends to be an awkward bulging shape, for maximum thickness is required at the quarters; but Maillart, the Swiss engineer, developed it into a thing of beauty. His variations on the theme are illustrated on pages 106 to 111.

Suspension cables are like arches in that their reactions at the abutments are inclined, not vertical. But they are reversed arches, for they pull at their abutments rather than push, and the cables or chains are wholly in tension whereas arch ribs are wholly in compression. Therefore the supporting members of suspension bridges must be composed of material that is not only flexible, but strong in tension: twisted vines or rope (page 32), metal chains (page 55) or steel wire (page 58).

Stone has an importance far beyond its limited use today, for it was in stone that the building of bridges first became a conscious art, and it is therefore stone that, for better or worse, has determined many of today's attitudes toward the esthetics of bridge design.

A slab of stone is an unlikely medium for a horizontal beam, for it is really strong only in compression whereas a beam must also be strong in tension if it is not to crack and break at the middle. Yet the beam rather than the arch was the first thought in stone construction—whether of buildings or of bridges—for it was the easier principle and familiar to any people who had dealt with wood.

When the Romans gave the arch to western civilization they gave the effective masonry bridge, for it is in the arch that the compressive strength of stone or brick comes into its own. The Roman pattern was simple enough: semicircular arches, rarely wider than 80 feet, supported by thick piers (usually about a third of the span) that took the lateral thrust of adjacent arches as well as their weight and thereby made each arch completely independent of its fellows. Piers were protected on the upstream side by pointed cutwaters and often lightened by small arched openings. The stones were dressed with utmost precision, and often laid up without mortar.

Most Roman bridges were devoid of ornament other than the strong moldings that marked the line of the roadway, plus the inherently decorative quality of fine stonework itself. Some of the urban Italian bridges, however, were embellished with projecting pillars or pilastered niches as well as elaborate cornices. Like their Renaissance and post-Renaissance imitations, these were better as buildings than as bridges, for the extraneous vertical accents tended to disrupt the sense of unity of purpose and continuity of line that is the very essence of a fine bridge.

Eastleach Martin Bridge over the Leach, England. Date unknown.

England has numerous slab or "clapper" bridges, some of which date from pre-Roman times while others seem to be less than three hundred years old.

The Greeks developed stone post-and-lintel construction to a peak of refinement in their temples, but showed little interest in bridges for they were a seafaring people and their waterways were their highways. The boldest stone slab bridges were built by the Chinese: the Cyclops themselves would have boasted of the fabulous 70-foot clear spans of the thirteenth-century Lo-yang Bridge in Fu-kien.

Pons Augustus (Ponte di Augusto), Rimini, Italy.
20 B. C. Spans of 28 and 23 feet.

Best preserved of the famous Roman bridges in Italy, this is characteristic in its nobility of proportion and its exemplary workmanship as well as in such specific features as the semicircular arches, uneven in number, the thick piers and the strong cornice with plain, round-topped parapet above.

But only the Ponte Rotto and the Sant' Angelo, both in Rome, were as elaborately ornamented. The cornice and its supporting modillions emphasize the continuity of the roadway even while the pilastered niches with their entablatures and pediments tend to disrupt it, though time has softened their outlines.

Beloved of Palladio (see page 20), this bridge, through his influence, has been imitated in many parts of the world

The diagram illustrates a popular type of Roman bridge, executed with particular dignity in Spain. The small arches over the angular cutwaters serve the practical purpose of lightening the massive piers and allowing additional passage for flood water; they also provide a counterpoint to the rhythm of the main arches. Beauty is sought and found in refinement of structure, and there is no recourse to extraneous ornament.

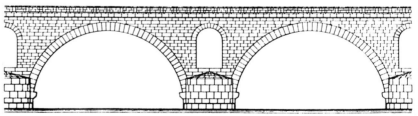

ROM, TIBERBRÜCKE, 60 V. ZW.

European society was so disorganized after the fall of the Roman Empire that bridges must have seemed neither necessary nor desirable, even if the art of their construction had been understood. The old Roman bridges were ravaged by war and neglect, and nothing new worth mentioning was built until the twelfth and thirteenth centuries, when the art was revived by groups of monks, Pontist Friars, who charged themselves with the assistance of travelers and pilgrims, particularly through the construction of bridges. The famous bridge at Avignon (below) was their work, also the original London Bridge (opposite). Later bridges, often heavily fortified, were built not by religious orders, nor often by feudal lords, but by the increasingly powerful cities.

Medieval bridges were crudely built as compared with their Roman ancestors, and some have little more than quaintness to recommend them. But by and large, medieval builders compensated for their technical shortcomings by the fresh, sure intuition with which they approached their ancient structural system of high-curving barrel vaults and massive piers. The very rudeness of their stonework is more lively than the exquisitely dressed, painstakingly coursed masonry of the Romans. And when they achieved the majestic simplicity of Newby Bridge (page 17) or the famous arch at Lucca (page 25), their shapes seemed to flower in beauty from some special awareness of the unity of material, structure and form.

The aspiration toward soaring lightness that was manifest in the great Gothic cathedrals was reflected in contemporary bridges, to such extent that daring occasionally exceeded ability, but curiously enough, the basic principle of cathedral construction—the concentration of arch weight and arch thrust upon isolated points of support—was not applied to bridges. Until the advent of Perronet (page 19) stone bridges were to rely for their strength upon brute mass alone.

Pont d'Avignon, over the Rhône at Avignon, France, 12th century. Built by St. Bénèzet, leader of the Pontist Friars in France.

This was the first of the great medieval bridges. Only four of the arches remain, yet they are sufficient indication of the special character of the bridge as distinguished from its Roman precedents. The old Roman discipline has gone, along with the fine workmanship, but new values appear in the flattened curve of the arches and the off-center accent of the little chapel that is rooted in one of the great piers. There is no dominant axis, no interruption of the rhythmic flow of the bridge by emphasis upon its center or its ends.

The apparent casualness of the composition is deceptive, for here, conscious or unconscious, is evidence of a new will to form.

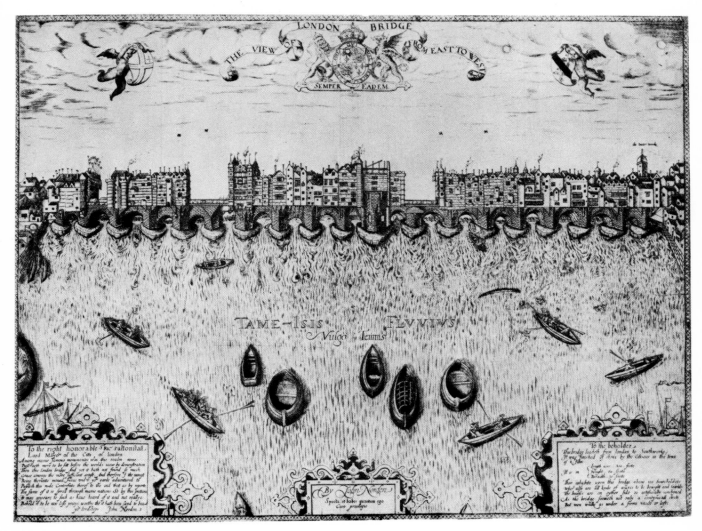

Old London Bridge. Started in 1176 by Peter Colechurch, a Benedictine monk who belonged to one of the famous bridge-building orders of the time; completed in 1209; replaced in 1824.

This first stone bridge over the Thames was crude indeed as compared with the work of ancient Rome, or even of contemporary France. The pointed arches, a medieval invention, exert less outward thrust than semicircular arches, but the advantage was not exploited. The narrowness of the irregular spans together with the great breadth of the piers and their even broader cutwaters made the bridge into an almost impassable dam.

This print of 1594, drawn by one John Norden, bears the following legend: "There inhabit upon this bridge about 100 householders where also are all kinds of wares to be bought and sold; the houses are on either side so artificially combined as the bridge seemeth not only a continuous street but men walk as under a firm vault or loft."

Great Haywood Bridge, Staffordshire, England.

The medieval bridge pictured here is very similar in design to the one illustrated on the opposite page. The roadway is so narrow that the triangular niches over the cutwaters offer useful shelter from oncoming traffic.

The provision of cutwaters on both faces of a bridge was a medieval innovation, technically preferable to the Roman practice of using cutwaters only on the upstream side.

Newby Bridge, Lancastershire, near Lake Windermere, England.

One of the most satisfactory of many handsome medieval English bridges, this is typical in its disregard for the line of the roadway. The parapet is continuous with the spandrel walls and even follows the sharp diagonals of the cutwaters, thus making the bridge into a rhythmic alternation of just two elements—rounded arches and elongated, projecting piers. The lively surface of the rude stonework is an excellent foil for the boldly defined shapes.

The level of technical achievement is thoroughly unremarkable, yet the expression of the Roman principle of massive piers and independent arches is as appropriate as any pattern developed by the Romans themselves.

Santa Trinitá Bridge, Florence. Designed by Michelangelo Buonarroti; built by
Bartolomeo Ammanati between 1566 and 1569; destroyed in World War II.
  With its flattened, subtly curved arches, its exquisite proportions, its air
of certainty and restraint, this was surely the most beautiful and original of all
Renaissance bridges.

For many centuries the bridge had been basically Roman. The Gothic cathedral builders had not cared to apply their revolutionary techniques of masonry construction —the ribbed vault and the flying buttress—to the humbler problem of the bridge. Renaissance architects had flattened and refined the semicircular Roman arch, but the structural principle itself remained unchanged until Perronet, French master builder of the eighteenth century and one of the world's first professional engineers, happened on the idea of *interdependent arches*.

The idea of transferring the thrust of an arch beyond its immediate vertical supports was not new, for it had been the principle of the Gothic flying buttress; but Perronet was the first to apply it to bridge design. By using his piers to take only vertical loads, and letting the thrust of the arches carry over from one to the next until it was met at the ends by strong abutments, he was able to reduce the thickness of his piers to less than half the usual ratio.

Perronet immediately realized the esthetic consequences of his technical innovation. The flattened arches now assumed a life of their own, continuous from one abutment to the other, and quite separate from the slim piers that raised them into the air. His Neuilly Bridge (below) in particular, with the lean, leaping curves of its arches and the long unbroken lines of its cornice and parapet, was a classic statement of the idea of continuous structure. As such, it is interesting to compare with today's continuous beams of steel or reinforced concrete.

Bridge over the Seine at Neuilly, near Paris. Built 1768-74; demolished in 1938.
Jean Rodolphe Perronet, engineer. Five 120-foot spans; piers 13 feet thick.
The elliptical arches were splayed out to become segmental arches at the face.

If any one person could be held responsible for the split of engineering from architecture it would be Palladio, the celebrated mid-sixteenth-century Italian architect. This he accomplished less by merit than by demerit, for as prime mover in the revival of ancient Roman glories he was the first and most influential representative of the architect as reviver and picture-maker rather than as builder. That was not too pernicious as long as the conditions of living and building remained fairly static, as they did through the eighteenth century, and as a matter of fact a multitude of fine buildings all over the western world must be ascribed to the Palladian influence; but it was an attitude that gave the architect no encouragement to face squarely the new problems and possibilities of construction that came with the machine. A separate profession had to be developed to meet the emergency, an engineering profession.

It is therefore interesting to compare the work of Perronet, the first great engineer, with the bridge designs of Palladio, the man who helped to make him possible. Whereas Perronet's bridges are bridges, Palladio's are self-contained architectural exercises. Their function of carrying a continuous roadway over a stream is quite incidental, and the water beneath seems a fortunate accident, a delightful mirror, rather than their reason for existence. This is obvious in the Palladian type that is embellished with shops and arcades, and it is more subtly evident in his adaptation of the Pons Augustus.

Palladio's interest in bridges was not limited to a revival and development of the Roman masonry style, though that was where he had his great importance. He was also the first to be concerned with the possibility of wooden trusses, a fact that suggests that he himself might have welcomed the opportunities of steel and reinforced concrete without the qualms that many centuries later beset his self-appointed disciples

Bridge design by Palladio, adapted from the ancient Pons Augustus at Rimini, shown on page 13.

Design by Palladio for a stone bridge carrying three separate footways and six rows of shops.

Pulteney Bridge over the Avon, Bath, England. 18th century.
The Palladian style flourished with exceptional grace in England, whence it came to us as "Georgian," and the finest Palladian bridges were built not in Italy but in England. This example differs from others in its understatement, and seems to carry the city over the water without itself claiming any excessive importance.

Pont du Gard, the Roman aqueduct at Nîmes, France. 14 A.D. Spans up to 80 feet.
 Semicircular arches in three tiers lift the aqueduct 155 feet above the stream.
Mortar was used only in the top arcade, for the stones of the lower tiers were so
precisely cut and fitted as to require neither mortar nor iron clamps. The colossal
aqueduct is so powerful in outline and proportion that it deserves the immortality
assured by its substantial construction.

Landwasser Viaduct for the Rhaetian Railway near Filisur, Canton Grisons, Switzerland. c. 1904.

Elongated arches and curved plan are brought together in a design of bold simplicity.

When a stone bridge must be both high and long, the problem is complicated by the difficulty of spanning great distances with that material. Arches are then necessarily either ranged in tiers, as in the Pont du Gard, or made uniformly tall and slim, as in this famous Swiss example, or given varied spans. Since this last procedure usually means that arches spring from different levels, it rarely leads to a satisfactory appearance.

23 **STONE**

The history of bridge building is largely a story of man's willful pursuit of lightness and his eventual triumph over inert mass. Perronet's interdependent arches were an important step forward. Another was the development of the open spandrel.

As long as builders knew only the Roman semicircular arch with solid spandrels (i.e., the walls between supporting vault and deck), a large single-span bridge had to be extremely high and, unless it were a peaked "camel's back," extremely heavy and massive. The challenge was thus not only to develop a flatter arch, but to lighten the weight of the bridge by opening up the spandrel walls.

Spandrel arches meant a gain in intelligibility as well as in lightness, for they threw emphasis upon the decisive structural importance of the main vault, and differentiated it from its burden. Today's open-spandrel arches of reinforced concrete are not a separate phenomenon, but an advanced stage of a development that started in Europe in the fourteenth century, though the Chinese had solved these problems with great elegance (see page 26) more than seven hundred years before.

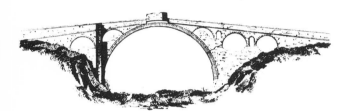

Céret Bridge over the Tech, France. 1321-39. 147-foot span.

The semicircular arch was in the Roman-medieval tradition, but the small arched openings in the spandrel walls represented a new thought in European bridge building.

The original outlines were blurred by repairs and the chapel over the crown, a characteristic medieval feature, was already ruined when this engraving was made in 1809.

Cabin John Aqueduct, in Maryland, near Washington, D. C. 1864. Built by General Meigs. 218-foot span.

The boldest stone arch in the United States and for forty years the broadest in the world, it is commendable for its forthright design as well as for its unusual dimensions. The flat top, accentuated by strong moldings, is less graceful over a great single arch than the more usual curve, but in an aqueduct it was mandatory. No attempt was made to lighten the massive stonework above the low-curving arc of the vault.

This is one of very few important stone bridges in this country. There are some charming miniature highway bridges, often dating from colonial times, and some fine early railroad viaducts, notably those designed by Benjamin Latrobe (the Younger) for the B. & O. Railroad; but American conditions favored bridges of wood and, later, of metal.

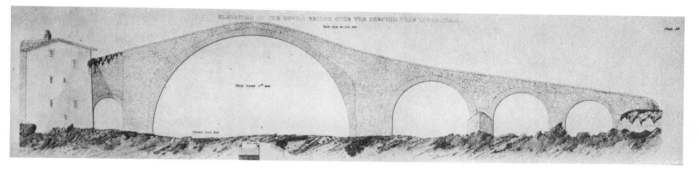

Ponte della Maddalena, over the Serchio near Lucca, Italy. 14th century. 120-foot span.

From the low river banks the road runs steeply up and over the great semicircular arch and the smaller side arches, making a "camel's back." Note the lively asymmetrical composition.

Brig 'a Doon, Alloway, Scotland.

This fine medieval bridge, beloved of Robert Burns, gains in decision through the crisp molding about its arch ring. Compare with the Swedish arch of reinforced concrete that is shown on page 91

An-chi Bridge at Chao Chou, Hopei, China. Built by Li Ch'un during the
Sui Dynasty, 590-616 A.D. 117-foot span.

This is the oldest open-spandrel bridge in the world. The low-rising arch ring,
a segment of a circle, is brought into sharp relief by the introduction of arched
openings in the spandrel walls. These serve not only to lighten the bridge but to
differentiate cause from effect, i.e., supporting arch from supported roadway.

The structure is phrased with such logic and grace, such acute awareness of its own
nature, that it makes most western bridges seem heavy and inarticulate by contrast.

Dolau-herian Bridge over the Towy near Llandovery, Wales. Designed by William Edwards (1719-89), who in 1755 built the similar Pontypridd Bridge in South Wales.

Edwards found through sad experience that his broad flat arches were feasible only when the pressure of the haunches was relieved by spandrel openings. These cylindrical perforations do nothing to differentiate the basic structural elements of the bridge, but they contribute a great deal to the beauty of the swelling arch.

Bridge in the Fenway, Boston. 1880-81. Henry Hobson Richardson, architect.
 The most beautiful bridges are not necessarily the most daring. Designed by a
famous pioneer of modern architecture, the Fenway bridge is frankly massive. With
a fine eye for texture, pattern and scale, the architect has laid up the great
blocks of stone in freely curving shapes that follow no historical precedent.
If the bridge looks at once medieval and modern; it is because Richardson,
here as in his buildings, used Romanesque masonry as his structural point of
departure, though not as an arbiter of specific form.

Overpass near Eisenberg, Germany. c. 1937.
For the *Reichsautobahn*: Karl Schaechterle
and Fritz Leonhardt, chief engineers;
Paul Bonatz, architect.

A bit dry and mechanical, this is
nevertheless one of the most handsome of
recent stone bridges. The spandrel arches
march off over the abutments without
break, and the graceful arcade is a superb
foil for the tapered curves of the main arch
and the sharp horizontal of the deck.

Stone construction is so laborious and time-consuming that it is justly considered an absurd and costly anachronism in a day when skilled labor comes high, while steel and reinforced concrete, true machine-age materials, are relatively cheap. Any contemporary American bridge that purports to be stone should therefore be regarded with suspicion, for close inspection will usually show that the stone is only a thin layer of deception applied to a structure of reinforced concrete.

Conditions were different in Nazi Germany. Short of steel and well provided with highly skilled masons, the Germans found stone bridges a not excessive luxury, and built many in connection with the impressive network of military highways that was their *Reichsautobahn*. But their reversion to traditional stonework cannot be fully justified on rational grounds. It must be attributed not only to their shortage of steel, but to their craving for self-glorification through familiar symbols of power and immortality.

In these German bridges stone was confined to actual vaults and visible surfaces, while the core was concrete. The stone skin, protective and ornamental, was also in a sense structural, for it was not an extraneous addition, but integral with the concrete beneath. Craftsmanship was remarkably good, but the over-all design, though superficially clean, was pallid and affected as compared with the best work of the past. The old vigor eluded the modern revivalists. Indeed, the German experiment proved rather conclusively that stone has run its splendid course as a bridge-building material.

Kok-su Bridge, China.
A log crib serves as intermediate pier for log spans in this primitive bridge, a type still built in the American backwoods.

It takes little thought and small skill to bridge a narrow stream with a felled tree, but the builder in wood—of whatever time or place—has needed considerable ingenuity to attain spans beyond the length or strength of any single timber. He has met the challenge in various ways. Sometimes he has shortened his span by projecting the vertical elements themselves out over the water, using some form of corbeling, bracketing or cantilevering. If he has been familiar with masonry construction, he has often bound his short timbers together to fashion great arches. Through trial and error he has also learned to join pieces of wood in triangles, forming strong rigid trusses. And from ancient times he has used plant material in the special form of twisted vines, bamboo splits or hemp rope, as cable for suspension bridges.

Like iron and steel, and unlike stone, wood is strong both in tension and in compression. Rather like iron and steel, and very unlike reinforced concrete, it must be dealt with as separate pieces. Obviously, then, it is generally suited to the same types of construction that are propitious in metal, and it is not surprising that the truss should have had its early development in wood, only later being translated into iron.

Old wooden truss, Tennessee.
The simplest type of truss is a triangle, from the apex of which a "king post" is hung to stiffen the horizontal beam at its bending point.

"New Railroad Bridge at Portage, New York." Mid-19th century. Wooden piecework.

Wandipore Bridge, Bhutan, between India and Tibet.
Great timbers are corbelled out toward each other from massive abutments and the narrowed interval finally capped with a light beam. This is the prototype of the modern cantilever bridge with "suspended span," such as those illustrated on pages 77 and 79.
Thomas Pope, who drew this picture for his book *Bridge Architecture* and used it as inspiration for his "flying lever" project, shown on page 36, said that the main span of the bridge was 112 feet, and that it was built in the seventeenth century.
Many bridges of this type were built in India and China.

Native cable bridge in Colombia, South America. From an old French print.

Bridge over the Min River, Szechwan Province, China.
Built since 1935. Total length of 1800 feet.

This multi-span footbridge with its twisted ropes of split bamboo follows ancient structural principles. The Chinese have long been adept with cable bridges, using bamboo rope or, even before the seventh century, iron chains, and they have always laid the deck directly on the curving cables.

The little gable roofs poised over the tapered towers protect the heavy timbers from damage by weather. Through the rhythmic repetition of these spare and handsome towers the bridge gains extraordinary distinction. One looks forward to the possibility of continuous-cable multi-span bridges in the contemporary American terms of flexible steel towers, steel wire cable and suspended level roadway.

Bridge south of Yunnan-fu, Yunnan Province, China.
    Here is perfection of proportion, and subtle variation of the broad flat rectangle
that is the dominant theme of the composition.

The "Burr-arch" truss.

Below is a half-section of the first example, a bridge over the Hudson at Waterford, New York, built in 1804 by Theodore Burr. Patented in 1817, the system was used for a majority of our covered bridges.

At the right is a view through a relatively late version at Rushville, Indiana, showing the light arch that supplemented the trusses.

The Town lattice truss.

Patented in 1820 by Ithiel Town, this truss needed no unusual lumber sizes and little tedious framing, and so became very popular. The diagonal web, continuous over river piers, took more gracefully to the gable roof than did the arch-and-truss combinations. This bridge, over the Connecticut at Orford, New Hampshire, was destroyed by flood in 1936.

Covered bridges can be beautiful, but they are self-contained, inarticulate, more like barns than bridges. The excitement of a daring leap is absent, for one cannot witness the spring. The cover is not for picturesque effect, but to protect structural timbers and their vulnerable joints from rain and snow. The roof also serves as cross-bracing. Most American wooden bridges were so efficiently covered over that their external appearance is quite independent of the mystery of their structure. Design became primarily an exercise in the just proportioning of roofs and walls and openings, the expression of portals, and the sympathetic relationship of bridge to landscape.

Covered bridges have been popular in many heavily forested countries, particularly in Switzerland, where some of the boldest in history were built in the eighteenth century by the famous Grubenmann brothers. But the richest development took place in the United States during the first half of the last century.

These bridges may look like barns, but their construction is, of course, far more complicated. Early American examples, like their European forebears, were usually pragmatic combinations of arches and trusses, for the truss was not yet fully understood and builders found it expedient to lean on the familiar principle of the arch. Wernwag's Colossus (see page 37), the longest wooden span of all time and possibly the most beautiful, was a truss-strengthened arch, but most builders used Burr's system of arch-strengthened trusses. The awkward "Burr-arch" truss was rivalled in popularity after 1820 by Town's lattice, a true truss in which the arch played no part. Transition to metal started in the forties with the use of iron rods as tension members in wooden trusses and interest soon shifted almost entirely to the new material.

West Hill Bridge, Montgomery, Vermont.
Photo by Edmund H. Royce from *The Covered Bridge* by H. W.
Congdon. Alfred A. Knopf, N. Y., 1946.

Let the broad arc the spacious Hudson stride
And span Columbia's rivers far more wide;
Convince the world America begins
To foster Arts, the ancient work of kings.
Stupendous plan! which none before e'er found,
That half an arc should stand upon the ground,
Without support while building, or a rest;
This caus'd the theorist's rage and sceptic's jest.

Thomas Pope's proposal for a "flying lever" bridge and his fash-
ionable couplets are from his *Bridge Architecture* of 1811, the
first of such treatises to be published in the United States. Pope's
optimistic span was to be of wood. Arched in form, it was yet a
cantilever-beam in principle, with the "flying levers" projected from
great masonry abutments, fitted out on the New York side as
apartments. He made a ⅜-inch scale model of half such a bridge
(of 1800-foot span) and according to witnesses the unsupported
arm, 50 feet long, took a 10-ton weight.

Below are structural details of a slightly different version of the
bridge. A fine flourish marks the spring of the intrepid cantilever,
and the chaste pyramidal abutments are an early instance of the
very satisfactory role that the freely interpreted Greek and Egyptian
styles were to play in the bridge architecture of the next few
decades. Pope, who described himself as an architect and landscape
gardener, was much concerned with the appearance of bridges,
and an avowed believer in "mechanical beauty."

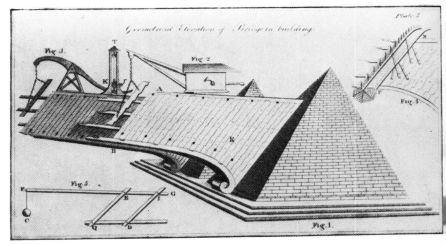

The Colossus, over the Schuylkill at Fairmount, Philadelphia. 1812; burned in 1838. Louis Wernwag, architect. 340-foot span.

This now seems to have been the greatest span ever achieved in wood or stone, for Dr. Joseph Killer of Switzerland has recently demonstrated that the Grubenmanns' Wettingen Bridge of 1764, long credited as supreme, was 200 rather than 390 feet in length.

There were five parallel laminated arches, each built up of seven thicknesses of timber, each strengthened by trusses above. Between these trussed arches ran two carriageways and two footways.

Judging from this view of 1823, the Colossus was as elegant in appearance as it was bold in structure. Note the rare grace of the arch, the complementary curve that marks the line of the roadway, the fine proportions of the windows. The neo-classic portals are less convincing.

Private footbridge near Princeton, New Jersey. 1942.
Kenneth Kassler, architect; Kraemer Luks, engineer.

In this day of drearily unimaginative wooden bridges the freshness and delicacy of this design are doubly conspicuous.

The tapered curves of the fixed arches are a lucid expression of the transfer of weight and thrust to the concrete abutments, and the relationship of these flattened arches to the much gentler curve of the footway is remarkably easy and graceful.

Temporary bridge for sight-seeing trains at the Züka Exhibition, Zurich, Switzerland. 1947. W. Stäubli, engineer; F. O. Kälin, consultant. Three-hinged arch of laminated wood.

Permanent arched bridges of glued and laminated wooden strips were built as long ago as 1907 by Otto Hetzer, a Swiss, and this exhibition bridge followed his system of construction. It is illustrated here less for its intrinsic merit than for its suggestion of future possibilities.

The bent and bulging ribs are no affectation, for their shape is a direct reflection of the static forces at work in a three-hinged arch. Rib edges become jagged as the bonded layers of wood decrease in number toward the hinges.

The picture shows the bridge under construction, before the outline of the arches was confused by the introduction of a second, lower deck.

A century or so ago wood lost its reputation as a suitable material for fine bridges. On every count—except, occasionally, that of initial cost—it seemed hopelessly inferior to the new materials offered by the Machine Age. Where its cheapness was a conclusive argument in favor of its use, the result was usually just a humble trestle or a bulky, awkward truss, best relegated to as remote a location as possible.

After this long period of neglect, wood is beginning to regain its lost dignity. Through the application of science, stimulated by the wartime shortage of steel, wood suddenly becomes a modern material. New methods of treating timber give it promise of relative permanence; the new metal ring connectors offer an easy and efficient substitute for the ancient and laborious practice of framing one timber into another in complex, vulnerable joints; and the new methods of lamination, of plastic-bonding short pieces of wood together to form huge monolithic beams and arches, offer interesting possibilities for new structure and new form.

A revival of the art of wooden bridge building seems due. Wood may never compete seriously with steel or reinforced concrete for long spans, but for small, light, or semi-permanent bridges its new potentialities for efficiency—and delight—are yet to be seriously explored.

Thomas Telford's project of 1801 for a "cast-iron bridge, consisting of a single arch 600 feet in the span, and calculated to supply the place of the present London Bridge."

The structural principle is that of the stone arch, for Telford planned to use Paine's system and build up his vault of small pierced blocks; but the transparent filigree, seemingly without weight or substance, is an imaginative interpretation of the special nature of the new material.

This elegant arch with its unprecedented span was unfortunately never executed, not because difficulty was foreseen in its construction, but because of its high ramped approaches, unwieldy in the city plan, and because of the general uncertainty of the times.

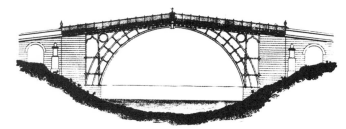

Coalbrookdale Bridge over the Severn, England. 1775-79. Designed by Thomas F. Pritchard, architect, for and with Abraham Darby III and John Wilkinson. 100-foot span.

This first iron arch is still in good condition, though the pressure of the earth behind the abutments has pointed what was originally a semicircular arch. The five separate arch ribs were cast in full halves at a nearby foundry—a construction very different from that of Paine's invention.

Not until 1836 was an iron bridge built in the United States.

Sunderland Bridge over the Wear, England. 1793-96. Rowland Burdon, builder. 236-foot span.

In 1790 Tom Paine had set up a successful experimental 110-foot arch in Paddington Green, London, placing it on exhibition with a shilling entrance fee, but the Sunderland Bridge was the first actual example of his ideas, and incorporated material from his experimental arch.

The abutments and the curve of the roadway are singularly awkward, though the arch is graceful enough. Each rib was built up of 105 of the cast-iron panels that are shown in detail at the foot of the page. Iron hoops filled the gap between arch and deck.

Component parts of the Sunderland arch ribs, from a contemporary engraving:
A. Side view of a block, about 5 feet high.
B. End view of a block.
C. One of the wrought-iron bars that join the blocks to form a rib.
D. One of the screw bolts.
E. Long view of one of the tubes used to unite the ribs horizontally.
F. End of a tube.
G. Four blocks united to form part of two adjacent ribs.

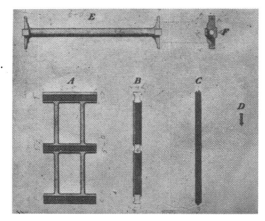

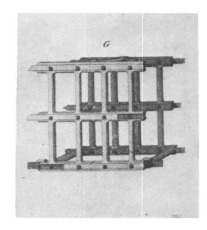

The great era of iron and steel was the nineteenth century, when most of metal's magic possibilities were explored and the very soul of the material revealed. Today we perfect and adapt, often with great skill, and occasionally we use our wealth of scientific knowledge as a basis for invention, but the original creative gusto seems somewhere to have been lost. Perhaps a material that is young and fresh is most stimulating to men's imaginations, closest to their hearts.

A full understanding of the new material was not immediate. When iron first appeared on the European scene as a likely structural medium for bridges, in eighteenth-century England, the impulse was to treat it like stone. When a material is so new that its own individual nature is not yet understood, the usual tendency is to handle it in the same manner as more familiar materials. Some iron chain bridges were built, as early as 1741 (see page 55), but the other early iron spans were all arches, and generally assembled like stone vaults of small panels of cast iron (see opposite page). This type of construction—actually quite sensible in cast iron, a brittle material that takes best to direct loads—was invented in 1786 by Tom Paine, the extraordinary American who was later to turn to political philosophy.

At no time, however, was there any imitation of the superficial appearance of masonry, or any attempt to duplicate its weightiness. Quite the contrary, iron seems to have been welcomed from the beginning as an honorable material, capable of a new and startling beauty of its own, and the transition from stone to metal, from mass to line, was accomplished with a minimum of esthetic fumbling. Masonry remained, of course, as piers, abutments and towers, and until 1850 or so the engineers, with or without architectural assistance, seem to have well understood the importance of shaping it into bold, clean-surfaced masses as a foil for their spidery ironwork.

The seventies and eighties saw the development of the modern bridge engineer, product of standardized scientific training. In the same period came the introduction and acceptance of steel as a material far stronger and more adaptable than iron. Exploiting new alloys, metal spans have become constantly longer, lighter.

But the gains of the last sixty years have been more quantitative than qualitative. Could it be that we are too intimidated by our science to preserve the courage of our intuitions?

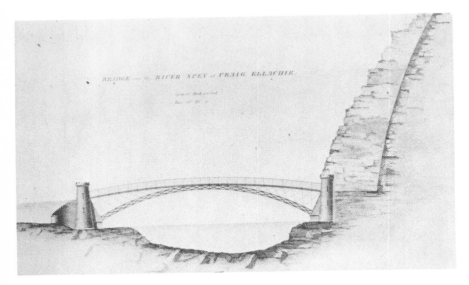

Craig Ellachie Bridge over the Spey, Banffshire, Scotland. 1813. Thomas Telford, engineer. 150-foot span.

More than other engineers, the great Telford was interested in economy of material, and the beautiful bridge that he flung over the Spey is extraordinarily delicate.

With its trussed arch ribs and its trussed spandrels above, this is generally credited as the first modern metal arch. It is still in use.

Chepstow Bridge over the Wye, Monmouth County, England. c. 1800. John Rennie, engineer.

This is a very early and appealing example of the iron multi-arch bridge. Like an attenuated spider web the arches stretch from one shore to the other, continuous beneath the long sweeping curve of the roadway and the light railings. The masonry is kept low, definitely subordinate. Compare with the Perronet bridge illustrated on page 19, and contrast with the Eads Bridge pictured below.

Eads Bridge over the Mississippi at St. Louis. 1868-74. James B. Eads, engineer. Spans of 502, 520 and 502 feet.

Technically the bridge was a great triumph, for its arches were of record span and it marked the first use of tubular structural members, of big pneumatic pier-caissons, and of extensive steel.

Eads' vigorous design is well illustrated by this rare early photograph.

The alternation of flat trussed arches with substantial masonry piers established an American design formula that is still with us today, frequently under the most unlikely of circumstances.

Viaduct over the Truyère at Garabit, France. 1884. Charles Eiffel, engineer.
Two-hinged arch of 545-foot span.

Like Eiffel's famous Tower in Paris, his Garabit Viaduct is an early triumph
of French engineering. Its great crescent arch asserts the concentration of forces
at the two abutment hinges, and the powerful outlines of the whole are compli-
mented by the lacy trusswork.

Yet the design tends to fall apart into its separate elements—arch, deck and
parading piers. Compare it in this respect with the magnificently single-minded
bridge that was built contemporaneously over the Firth of Forth (page 77).

Rainbow Bridge, Niagara Falls, New York. 1941. Waddell & Hardesty, engineers; Aymar Embury II, architect. Hingeless arch of 950-foot span.

Designed as a "set piece" with precise boundaries and highly mannered detail, this plate girder arch with slim spandrel posts is nevertheless one of the finest American examples of its kind.

Marble Canyon Bridge over the Colorado, Arizona. 1928. By the Arizona State Highway Department: L. C. Lashmet, engineer. Two-hinged braced-spandrel arch of 616-foot span.

The bridge is less forceful as an arch and less expressive of its two hinges than the Garabit Viaduct shown on the preceding page, or the Bietschtal Viaduct illustrated below; but it is more expressive of its material than the plate girder arch shown above. It fits very comfortably between the canyon walls, and its extreme delicacy is particularly welcome in the giant-scaled landscape.

Bietschtal Viaduct near Brig, Switzerland. 1913. Adolf Herzog, engineer. Two-hinged arch of 311-foot span.

The transparent, sharply articulated, powerfully jabbing shapes carry a conviction that has nothing to do with prettiness. Regrettable, however, is the choice of points on the slant of the arch for the support of the underslung side trusses.

Belt Parkway Footbridge, Brooklyn, New York. 1939. Designed by Clarence C. Combs, New York City Parks Department. A three-hinged plate girder arch.

The gracious curve of the arch is pointed up by the radiating lines of girder stiffeners and railing posts. Even the faces of the abutments are inclined at a sympathetic angle. But the problem of how finally to straighten out these diagonals at the ends of the bridge finds no very happy solution.

The span looks heavier than its load of pedestrians would seem to justify.

Plate girder arches with a minimum of trussing are today generally considered the handsomest of all possible types of steel bridge, with the single exception of the suspension bridge. They can indeed be extremely elegant, but they seem to lack the vigor of the best trussed arches and are certainly less eloquent of the special nature of their material. A plate girder arch with metal spandrel posts (like the fine West Bridge on page 46) is, from a distance, almost impossible to differentiate from a reinforced concrete arch with reinforced concrete spandrel posts (such as the bridge shown on page 90, top).

Västerbron (West Bridge), Stockholm. See opposite page.

Proposed aluminum arch over the Canimar River near Matanzas, Cuba. Designed in 1946. O. H. Ammann, engineer. 600-foot span.

If this arch is built it will be the first large bridge of aluminum. It will be very light in weight as compared with a similar bridge of steel. It will also be very expensive. Yet neither outlines nor dimensions seem visibly affected by the unconventional choice of material.

In comparison with the Västerbron, the Ammann design has its weaknesses as well as its merits. The fixed arch curves clear of the horizontal deck girders with a decision lacking in the Swedish bridge, but the heavy rigid frames over the abutments are less fortunate, while the parallel incisions that adorn the concrete are a regrettable pseudo-modern cliché.

Västerbron (West Bridge), Stockholm. c. 1935. By the Stockholm Harbor Board:
Ernst Nilsson and S. Kasarnowsky, engineers. Fixed arches of 668 and 551 feet.
   Slender steel posts carry the deck equably over arches, abutments and side
slopes, and the effect of limitlessness is accentuated by the vertical stiffeners of
the deck girders, repeated over and over until they disappear in the dim distance.
   With a refreshing absence of histrionics the angular concrete abutments are
concisely tailored to meet the thrust of the tapered, smoothly welded arch ribs.

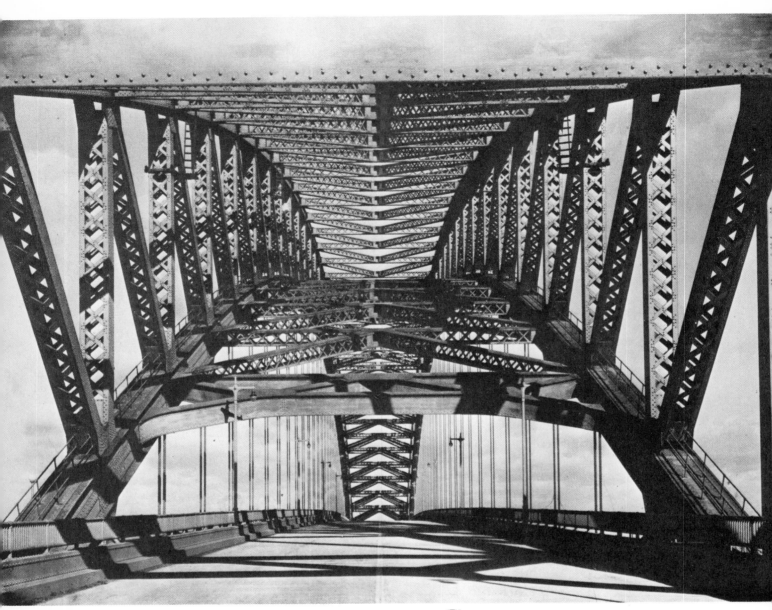

Bayonne Bridge, New York and New Jersey. See opposite page.

Bayonne Bridge over the Kill van Kull, New York and New Jersey. 1931. By the Port of New York Authority: O. H. Ammann, chief engineer. Two-hinged arch of 1652 feet.

The arch is magnificent, worthy of its fame as the longest in the world, and the fine-spun web of its trussing offers magic perspectives. But the design fails completely at the abutments, where the thrust of the giant arch is apparently met only by a light steel framework, an obvious impossibility that calls for explanation. Actually the weight and thrust of the arch is safely passed through its hinged lower ends to the massive piles of concrete at the base of the steel framework; but the engineers seem to have tried to disguise their hinged arch as a deep-ended fixed arch, a form hallowed by long association with stone construction, for they planned granite-faced towers that would look strong enough to take the non-existent thrust of the thickened arch ends. The idea of the stone facing was abandoned just as it was in the George Washington Bridge (page 59), built at the same time by the same Authority, but in both cases the indifferent framework of the towers, obviously not designed for display, remains as testimony to the original intention.

Compare this bridge with an earlier two-hinged trussed arch, the Garabit Viaduct, shown on page 43.

Alton Railroad Overpass, Mazonia, Illinois. 1939. By the
Illinois State Highway Department.
    Rolled steel sections make an unusually neat, though doubtless
expensive solution to the difficult problem of bracing a
overhead arch.

Bridge over the Connecticut at Orford, New Hampshire. 1937.
By the State of New Hampshire Highway Department. Two-hinged
tied arch of 425-foot span.
    This view along the roadway is less prepossessing than the side
view of the same bridge, shown below. The confusion of the bracing
was perhaps avoidable only at prohibitive additional cost, but there
could be no such excuse for the massive stepped-back parapets
at the entrances.

Bridge over the Connecticut at Orford, New Hampshire.
    Compare the outlines of this "bowstring" arch with the free
intersection of arch and roadway in the otherwise identical
bridge pictured on the opposite page.

Bridge over the Connecticut from Chesterfield, New Hampshire, to Brattleboro, Vermont. 1936-37. By the State of New Hampshire Highway Department. Two-hinged arch of 425-foot span.

The roadway intersects the arch at a level well chosen to flatter its curve, but the abutments might be more clearly expressed, better differentiated from the retaining walls behind.

The bridge is painted the color of young lettuce, a refreshing change from the usual blue-gray.

The designer of an overhead arch is confronted with all the usual problems of arched bridges plus two that are peculiar to his task. The first has to do with the overhead bracing, which must be kept as light and clean as possible. The second problem concerns the intersection of arch and roadway, which should take place at a point that will not only provide the requisite clearance above water and the needed convenience of approach, but will also maintain the visual integrity of both arch and deck through a just relationship of the two.

The ubiquitous "bowstring" arch, in which the roadway ties together the extreme ends of the arch ribs, seems to destroy both the character of the arch and the continuity of the roadway, but any other generalization is dangerous, for a relationship that is satisfactory in one instance may under other conditions be extremely clumsy.

Highway bridge over the Vilaine at La Roche-Bernard, France. 1912.
Daydé, engineer. Three-hinged arch of 656-foot span.

With a magnificent disregard for conventional canons of beauty, the lightly
drawn, somewhat ominous arch attains a splendor all its own. Perhaps only a
French engineer would have been capable of the gesture.

Compare this with the suave footbridge illustrated on the opposite page.
There is no question as to which is the prettier of the two, but there is also
little doubt about which is the more vigorous.

Footbridge over Lake Shore Drive, Chicago. 1940. By the Engineering
Division of the Chicago Park District: Ralph H. Burke, chief engineer.
Three-hinged arch of 187-foot span.

The attenuated arch and the long slow curve of the walk are brought
together with complete felicity, their ever-changing relationship measured by
the steady beat of the light steel sections that serve as hangers and posts.
The latticed overhead bracing is fine-scaled and unobtrusive; but the design
of abutments and approaches is mannered and heavy-handed—not to be
compared in quality with the span itself.

The arch is a matter of weight, gravity and pressure. As such, it is relatively passive, earthbound. The suspended cable reverses the arch curve and grows wings. Impatient of gravity, it achieves strength without apparent mass or weight. Substance seems transmuted to line, inert matter to naked energy.

Large suspension bridges attain a measure of esthetic value on a simple quantitative basis, for extremes of lightness, length and height are in themselves sufficient to arouse emotion. But that initial awe can only be sustained by quality—by a pleasing cable curve, by a just relationship between the main span and the smaller side spans, between the portion of the towers above the main deck and the portion below, and very importantly, by appropriate design of the towers themselves.

It is in this last respect that bridges most often fall short of perfection. The transition from the tower of stone to the flexible tower of steel completed the transition from mass to line and theoretically banished the last remnant of obeisance to the old masonry-born concepts of strength as weight, beauty as mass; but the implications involved in this change of material have not always been either welcomed or understood. Regretting the loss of the easy monumental possibilities of solid masonry and uncertain of esthetic substitutes appropriate to their new material, tower designers have tended to relapse into a helpless confusion of structural and ornamental or semi-ornamental forms.

Rope bridges have been built from time immemorial (see page 31), and iron chains were used for the purpose as early as the seventh century in the Orient, by 1741 in Europe; but it was an American, James Finley, who in 1801 first suspended a *level roadway* from his shore-to-shore cables, rather than laying the deck directly upon them, as had been customary; and a Finley-patent bridge of 1816 may have been the first to use wire cable rather than iron chains. The United States contributed little more until the middle of the century, but these inventions were immediately followed up in Europe. The English held to their chains of linked iron bars, and with them achieved spans that were sometimes miracles of lightness and grace, and all the more effective by contrast with the colossal masonry of the supporting towers. The French and the Swiss, particularly interested in wire, developed methods of spinning the cable in position and in 1834 achieved a record span of 810 feet at Fribourg, Switzerland. But leadership passed back to the United States in 1848, when Charles Ellet built a bridge of 1010-foot span over the Ohio at Wheeling, West Virginia; and thanks largely to the genius of the Roeblings, it has remained here ever since.

The great modern suspension bridge is an American phenomenon, encouraged by the peninsular sites of two major centers—New York and San Francisco. Foreign examples are relatively few, relatively small, though sometimes very handsome. And it is the only bridge type in which the United States excels in lightness, for extreme economy of material, therefore of bridge weight, is the *sine qua non* of such tremendous spans.

If suspension bridges look lighter than other types of steel bridge, it is because they *are* lighter. The supporting cable, being wholly in tension, takes full advantage of the fact that steel is far more efficiently used in tension than in compression. Thus it is in the suspension bridge that the nature of steel is most completely realized.

Winch Bridge over the Tees, England. 1741. 71-foot span.
  In this earliest European chain bridge, the flooring was laid
directly upon the cables. Beneath were diagonal chains that served
as wind bracing.

Menai Straits Bridge, Wales. Built 1819-24; recently widened.
Thomas Telford, engineer. 570-foot span.
    The flat curve of the iron bar-chains over the main span is admirable,
but the bluntly tapered towers are undistinguished and the support
of side spans by both iron hangers and masonry arches is
disconcertingly redundant.

55 **SUSPENSION CABLE**

The Avon Competition.

In 1829 the city fathers of Bristol selected the great Telford, who had just completed the Menai Bridge, as referee of a competition for the design of a suspension bridge over the Avon at Clifton. He judged all entries unsuitable, including the proposal of I. K. Brunel (a), which he declared too long in span. Asked to submit his own recommendation, Telford obliged with the fantasy shown at bottom (d). This was approved but not executed, and in 1830 another competition was held, this time with Telford as contestant. The winner (b) was the twenty-four-year Brunel, and the final result was his Clifton Bridge, illustrated on the opposite page.

Brunel's early designs for the Clifton Bridge.
a) Above is his fine project in the first competition: inverted cables below roadway for stiffening; Norman castles as anchorages; no towers. 916-foot span.
b) Below is the design on which work was commenced in 1836.

c) This design was submitted in the competition by one C. H. Capper, "engineer." A favorite motif of the time was the built-in medieval ruin, here at its most delightfully incongruous.

d) Telford's Gothic design for the Clifton Bridge shows that the early nineteenth-century engineers were not unaffected by the romantic fallacies then current among the architects. Even the chains were to be ornamented with fretwork.

Clifton Bridge (above and right) over the Avon, Bristol, England. I. K. Brunel, engineer. Begun 1836; completed after Brunel's death in 1859. 702-foot span.

This is by far the most beautiful of the early suspension bridges, and through its great height still one of the most spectacular. The cable threads across the void in a shallow curve over the thin line of the roadway, then dips straight from the towers to anchorages in the rocks behind. The towers themselves are magnificent. Their inspiration is Egyptian, but the masonry is so boldly and freely shaped in response to function and material as to seem inevitable. Their style is finally not Egypt's, but their own.

The executed bridge is a vast improvement over Brunel's original scheme, illustrated on the opposite page (b); superfluous side hangers have been omitted, side cables drawn taut, and the towers given a much more vigorously expressive shape.

Since work on the superstructure started only the year following Brunel's death, and since his son and biographer complained in 1870 that "no attempt has been made to complete the towers according to Mr. Brunel's architectural designs," it is perhaps a mistake to give Brunel entire credit for the simplification and refinement of the towers. But the quotation may refer merely to the omission of the cast-iron plates, decorated with Egyptian-style figure drawings showing various stages of construction work on the bridge, that Brunel had hoped to use as sheathing for the masonry.

Brooklyn Bridge, New York. 1869-83. Designed by John Roebling;
executed by his son, Washington A. Roebling. 1595-foot span.

In the Brooklyn Bridge, two materials of opposite nature are brought
together in harmony: granite, strong in compression, piled majestically
into the sky; steel wire, strong in tension, spun lightly through space.
Last and noblest of the great stone-towered suspension bridges, this was
also, in its fantastic boldness, its wealth of technical invention and
very particularly, in its use of *steel* wire, the prototype of the huge
bridges of the 1930s. The 135-foot clearance set the present standard
for bridges over navigable waters, and the diagonal storm-stays,
radiating down from the tower tops, are being reintroduced today as

a precaution against such a disaster as befell the Tacoma Bridge
(page 63). Here storm cables and vertical suspender cables make a
diaphanous web.

Suspension bridges generally look best when side openings are less
than half the length of the main span, and when the cables at mid-span
curve clear of the roadway. The Brooklyn Bridge meets neither condition,
for its span division is 930-1595-930, and its cables drop almost to the
bottom of its double deck. Yet these shortcomings are not particularly
disturbing, perhaps because of the strong curve taken by the
trussed roadway.

George Washington Bridge, New York. 1927-31.
For the Port of New York Authority:
O. H. Ammann, chief engineer; Cass Gilbert,
consulting architect. 3500-foot span.

As the river view above testifies, the
George Washington Bridge is remarkable not
only for its size, but for the excellence of its
proportions. Note the short side spans, the
shallow curve of the cables and the almost
incredible thinness of the roadway. The bridge is
unique in its lack of longitudinal stiffening trusses
or girders. It is stabilized instead by its own
great weight, for it is the heaviest single-span
suspension bridge ever built.

Thirsty for an appearance of orthodox
monumentality, the designers built up the steel
framework of their 635-foot towers with the idea
of casing it later in masonry. The mask was
omitted because of popular protest, but the
meaningless arches remain as testimony to the
original intention. Nor was appearance improved
by the addition of a top story over the cable
saddles, absent in the early photograph above.

Golden Gate Bridge, San Francisco. 1933-37. Joseph B. Strauss, chief engineer; O. H. Ammann, Leon S. Moisseiff and Charles Derleth, Jr., consulting engineers; Irving F. Morrow, consulting architect. 4200-foot span, longest in the world.

The bridge is fortunate in its colossal dimensions, its permanent coat of orange paint, and its spectacular surroundings, but the quality of its design is not commensurate with its size. The towers, looming 746 feet above water, are capricious in outline and detail, and take poorly to the relatively low placed, trussed roadway and the extraordinarily deep curve of the cables.

Chûte du Brûlé Bridge over the Gatineau River, Province of Quebec, Canada. 1938. Designed by the Dominion Bridge Company, Limited. 300-foot span.

Quiet, graceful towers and clean, purposeful lines make this probably the finest small suspension bridge in the Americas. The relatively deep stiffening girders are in the European tradition; compare with the Rodenkirchen bridge shown on page 62.

Bronx-Whitestone Bridge, New York. 1939. For the Triborough Bridge Authority: O. H. Ammann, chief engineer; Allston Dana, engineer of design; Aymar Embury II, architect. 2300-foot span.

In this first use of shallow plate girders as roadway stiffeners rather than the customary deep trusses, the ratio of girder depth to span was a mere 1/210. Since the Tacoma failure (page 63), the 11-foot plate girders have been reinforced above by trusses to a total depth of 25 feet, entailing complete loss of the original proportions. Diagonal storm stays have also been added, run from tower tops to roadway as in the Brooklyn Bridge.

The picture shows the bridge in its original condition, with a fine-spun elegance of outline and detail that was unique among modern suspension bridges, although the big and little arches of the tapered 377-foot rigid-frame towers are obviously something of an affectation.

Rodenkirchen Bridge over the Rhine, Cologne, Germany. 1938-41; destroyed in
World War II. For the *Reichsautobahn*: Karl Schaechterle, Fritz Leonhardt and A. Klönne,
engineers; Paul Bonatz, architect. 1244-foot span.

This largest European suspension bridge was no miracle of lightness as compared
with American achievements. Its beauty was wholly a matter of terse, highly
articulate structure and exquisite proportions. There was not one empty gesture or
superfluous word, and each smallest part was dignified by its coherent relationship
to the whole.

The low clearance permitted by the Rhine's small-scale shipping activities was
obviously a great advantage to the designers of this bridge.

Tacoma Narrows Bridge, Washington. 1940; ruptured by wind four months after completion. 2800-foot span.

This handsome bridge was modeled after the Bronx-Whitestone, but was longer in span and measured only 39 feet between cables. The roadway was stiffened only by 8-foot-deep plate girders, a ratio of girder depth to span (1/350) that has been exceeded only by the George Washington Bridge, where weight gives stability.

The distinguished committee that investigated the failure reported that "excessive vertical and torsional oscillations were made possible by the extraordinary degree of flexibility of the structure and its relatively small capacity to absorb dynamic forces." Many of the early suspension bridges suffered a similar fate.

The failure of the Tacoma Bridge was a shock and a challenge to American bridge engineers. The immediate response was unfortunate, for apprehensive engineers and public officials hurriedly retreated to the apparent safety of deep, ungainly stiffening trusses—the obvious, orthodox antidote to excessive flexibility in suspension bridges.

Yet the disaster may finally prove beneficial, for it has inspired a wealth of conscientious research and fresh creative thought, some of which now begins to bear fruit. Most of the new proposals have to do with ways in which proper aerodynamic design of the roadway—usually through some type of streamlining coupled with a system of road vents for free wind passage—will permit retention of its shallowness and flexibility. But the boldest proposal suggests a radical change in both the structure and the form of suspension bridges, for it discards the principle of flexibility. Instead, the wire hangers become the diagonal members of a *cable truss* of extraordinary lightness and stiffness.

Britannia Tubular Bridge for the Chester and Holyhead Railway, over the Menai Straits, Wales. 1846-50. Robert Stephenson, engineer; Francis Thompson, architect. Spans of 230, 460, 460 and 230 feet.

Trains run through twin wrought-iron beams, laid side by side, each a continuous rectangular tube of 1511 feet. The marble-faced towers, the central one of which is 230 feet high, were conceived as supports for auxiliary chains, but the hollow girders proved so strong in themselves that the cables were omitted.

This was the first great assertion of the flat beam in modern bridge building. The construction was revolutionary, but no attempt was made to recall more familiar structural forms, and no compromise was allowed to blur the decisive relationship of horizontal and vertical, metal and masonry. Even the Greco-Egyptian overtones of the towers seem unaffected and curiously harmonious; compare with Brunel's masterpiece at Clifton, shown on page 57.

This seems to have been one of the first major examples of the successful collaboration of a bridge engineer and an architect. Thompson was also the designer of numerous railway stations on the same line.

A standard design for underpasses below the German Autobahn. c. 1937. Karl Schaechterle and Fritz Leonhardt, chief engineers; Paul Bonatz, architect.

The plate girder bridge achieves elegance through refinement of structure. Vertical stiffeners divide the girder into flat rectangles similar in proportion to the opening itself, and the continuity of the roadway is stressed by light unaccented railings, and by the prolongation of the cantilevered sidewalk slab as a coping over the retaining walls.

Note that the seat of the girder is visible at either end, sharply distinguished from the retaining walls, and that the banks have been held back to allow the main lines of the bridge to come free.

A plate girder bridge is immobile and a bit dry as compared with an arch or a suspension bridge, and in long heavy spans it is likely to seem gross as compared with a fine-membered truss. But it is a good simple elementary form, orderly and restful, and at its best—shallow, cleanly drawn, crisply detailed—it is not only pleasantly unobtrusive but notably elegant.

It was not until about 1830, when the principles of arch and cable were already highly developed in metal, though the truss was still pragmatic, still wooden, that George Stephenson, a famous English engineer, first thought of building one of his railway bridges of flat solid-walled iron girders. The great Britannia Bridge, shown opposite, was the work of his son Robert. Its tubular structure is of course rather different from a plain plate girder, but the esthetic problems involved in its design were very similar. In spite of this illustrious ancestor the plate girder was long dismissed as a humbly utilitarian kind of construction, useful enough for out-of-the-way railroad viaducts but definitely unworthy of creative attention. Only in the last fifteen years or so, when its economy for small spans has made the steel plate girder universally popular for highway bridges and overpasses, have its potentialities for beauty begun to be recognized along with its practical advantages.

The girders need not invariably have a straight lower edge. When they are not divided into separate spans but made continuous over a number of openings, an economical procedure, then greatest strength is usually needed to meet stresses and strains concentrated over the vertical supports. The girders may merely be thickened at these points, and their under-edges kept flat, as in the German example pictured on the next page; or the extra strength may be provided through increased girder depth over the piers. In this latter case, the transition may be effected with technical propriety either by smooth curves or by sharp diagonals; the first look well under an up-curving roadway, the second beneath a flat deck. In either event the girder is logically brought straight to its end-pier or abutment, and thus differentiated from the true arch construction that it fortuitously resembles. Flowing curves or sharp bends are also in order in the rigid frame, and again the temptation to imitate true arches must be checked in the interest of esthetic integrity.

Because of its lack of structural drama the plate girder more than any other bridge type depends for success upon justice of proportions and perfection of detail. Sidewalks, railings and abutments assume decisive importance, and the quality of the whole is very much affected by the design of the piers: by their spacing, which determines the proportions of openings; and by their shaping, preferably as solid, quiet slabs of masonry, or as light, expressively contoured rigid frames of steel or reinforced concrete.

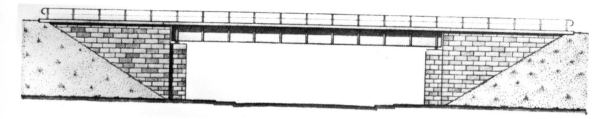

Bridge over the Freiberg Creek at Siebenlehen, Germany. 1938.
For the *Autobahn*: Karl Schaechterle and Fritz Leonhardt, chief engineers;
Paul Bonatz, architect.

The colossal masonry piers, flat-sided and slightly tapered, are
capped with rollers to allow for movement of the continuous girders in
response to temperature changes. Vertical stiffeners divide the girders
into up-ended rectangles, and the interval between stiffeners is the unit
that determines the location of sidewalk brackets and railing supports.

The engineers of the Nazi bureaucracy were at their best when
dealing with the sober problem of the plate girder and achieved some
extraordinarily handsome solutions, of which this is one of the finest.
Incidentally, the discussion of the design of plate girder bridges is
a particularly valuable chapter in the excellent book on bridge esthetics
written by this same Schaechterle and Leonhardt: *Die Gestaltung der
Brücke* (The Design of Bridges), published in Berlin in 1937.

Birdsong Creek Bridge, near Camden, Tennessee. 1942. By the
Tennessee Valley Authority. Main span of 103 feet.

Like most TVA bridges, this continuous plate girder is not
remarkably light in appearance, but most remarkably neat. Dominant
are the attenuated, well-differentiated horizontals of the steel
girder, the projected concrete coping, the low concrete parapet and
the metal railings.

This clarity is absent at either end, for girder seat, retaining walls
and parapet are brought together as one inarticulate mass of
concrete. The limits of the bridge are so sharply defined by these
terminal accents as to threaten one's sense of the roadway as
continuous.

Bridge over Fontana Reservoir for the Stecoah-Bryson City Road,
North Carolina. 1944. By the Tennessee Valley Authority.
Spans of 189, 228 and 189 feet.

Instead of simply thickening the continuous girders over the
piers to provide the extra strength needed at those points, the
designers have chosen to increase the depth of the girders, bringing
them down in straight diagonal lines that look very well under
the flat road-deck. The shadow cast by the cantilevered sidewalk
makes the bridge seem unusually lively and three-dimensional.

Now that Fontana Dam is completed, the tall piers are partially
submerged. The tilted roadway is not a photographic distortion,
for one end of the bridge is actually considerably higher than
the other.

Gowanus Elevated Parkway, Brooklyn, New York. 1941.
By the Triborough Bridge Authority.

Proud symbol of a new age, the highway cuts above its dreary
surroundings, its long slim legs withdrawn from chaos. These
rigid-frame supports would be better without their weakly drawn,
arbitrary arches, but otherwise they are remarkably clean
and powerful.

Seen from beneath, the structure of the boldly cantilevered
roadway is very expressive, inherently ornamental as it tapers up
and out from the longitudinal girders.

Old Alfred Road Overpass, on the Maine State Turnpike, near Biddeford, Maine. 1947. By the Maine Turnpike Authority. Howard, Needles, Tammen and Bergendoff, engineers. Continuous girder with spans of 36, 58, 58 and 36 feet.

American engineers have only recently begun to concern themselves seriously with the appropriate design of plate girders, but their products are already noticeably cleaner and more agreeable than they were ten years ago.

Here the deck is carried by the cantilevered arms of reinforced concrete piers. The piers themselves are set askew to parallel the divided highway beneath and thus to offer it a minimum of interference, physically and psychologically.

Saco River Bridge, on the Maine State Turnpike, near Saco, Maine. 1947. By the Maine Turnpike Authority: Howard, Needles, Tammen and Bergendoff, engineers. Continuous girder with two spans of 90 feet and four spans of 110 feet.

The bridge is run over the river in two separate halves to carry the divided highway of the Turnpike.

Brackets project from the girder wall to support coping and railing. Although the shaping of these projections seems willfully labored on close inspection, they make a merry effect as they run the length of the bridge.

North Chickamauga Creek Bridge, near Chattanooga, Tennessee. 1940.
By the Tennessee Valley Authority. Continuous girder with spans of 52, 94 and 52 feet;
bridge width of 30 feet.

Here the TVA engineers and architects have produced one of the finest small
bridges in the United States. Bold horizontals and diagonals are accentuated by the
regularly repeated verticals of the girder stiffeners. Perhaps it is this uncompromising
decision of line that makes the bridge more exciting than the softer, sweeter
design shown at the foot of the opposite page.

◀

Valley River Footbridge, Murphy, North Carolina. 1939. By the Tennessee Valley
Authority. Spans of 52, 78 and 52 feet.

The continuous girder curves down at the intermediate piers, harmonizing with the
much gentler curve of the walkway itself. The flat run to the abutments adds vigor
to grace and successfully cancels the incidental resemblance to arch construction.
Rivet heads pattern the surface of the slender girder, but the usual vertical stiffeners
are absent.

Proportions are excellent. Only the narrow little shelves that support the girder
ends are awkward; better to suppress them completely, as in the bridge pictured
above, or better still, give them more distinct expression, as in the German
underpass pictured on page 65.

Bridge over the Rio Malleco, Chile. 1886-89.
The spidery mesh of girders and towers is unmistakably of the last century, and probably the work of a French engineer.

The bridge is so delicate, so transparent, that it threads over the valley without seeming to disturb it.

Old iron truss, Tennessee.
When bridges such as this are replaced by a type of truss better suited to modern traffic, the new bridge rarely offers compensation for the loss of finely etched lines.
Here the main span is a Pratt-type truss, in which diagonals, pure tension members, were always iron rods, while verticals were either of wood or of iron. Invented in the United States in 1844, this was one of the earliest of scientifically designed trusses.

Pit River Bridge, over Shasta Reservoir, near Redding, California. 1941. Designed by the engineers of the U. S. Bureau of Reclamation.

This double-deck bridge is one of the handsomest trusses in this country. Before the development of the theory of the continuous beam, the unequal spans would have called for girders of varying depths.

There seems to be no good reason why run-of-the-mill overhead trusses cannot be comparably quiet and horizontal in design.

Nineteenth-century trusses had a gossamer quality that is rare today. Modern trusses must in actuality be heavier to take today's heavier loads, but their bulkiness must partly be ascribed to the use of relatively few, relatively large truss members, whereas nineteenth-century engineers preferred a close web of many light members, airy in appearance yet well defined in space as a semi-transparent geometric plane. Piers too, especially in France, were often of lacy trusswork.

The overhead or "through" truss is economical, therefore prevalent, but pleasant solutions are almost non-existent. The popularity of the lumpy hump-backed version is particularly regrettable, for an uneven upper edge generally looks nervous and clumsy, and complicates even further the complicated problem of overhead bracing. Nevertheless, the sky line of a fully developed cantilever truss with arched "suspended span," such as those shown on page 77 and 79, has an expressive vigor that can be quite magnificent.

Bridge over the Rhine at Neuwied, Germany.
c. 1934; destroyed in World War II.
Karl Schaechterle and Fritz Leonhardt,
chief engineers.
   This truss was dignified by the care for
outline and detail that is normally reserved for
more pretentious structures.
   All truss members were inclined at an angle
of 63 degrees. This absence of verticals made
for an unusually coherent pattern, intelligible
from every viewpoint. The smooth rigid-frame
portals were very neat, also the lattice
bracing above the roadway.

Goethals Bridge over the Arthur Kill at
Elizabeth, New Jersey. 1928. For the Port of
New York Authority: Waddell & Hardesty,
engineers.
   The internal confusion is typical of
an overhead truss with an uneven upper edge.
Compare with the orderly German bridge
illustrated above.

Bridge over the Sitter between Haggen and Stein, Switzerland. 1937.
Rudolf Dick, engineer. Continuous truss with spans up to 228 feet.
   Proportions are fantastic, for the tallest pier is 276 feet high, while the total width
of the single-lane roadway and the two sidewalks is only eleven feet. The deck
and its supporting trusses bulge out over two of the main piers to allow automobiles
to pass.
   The extraordinary lightness and laciness of the trussing is pleasantly reminiscent
of nineteenth-century French practice, but the sweeping high-crossed lines of the
lean, tapered legs make a shape definitely of our own day.

Bailey bridges of World War II. Invented by Donald C. Bailey of the British Ministry of Supply. Spans up to 240 feet.

Bridges are assembled of prefabricated, interchangeable trussed panels, each ten feet long and designed for handling by six men. Panels are pinned together on the ground, then shoved out over the water on rollers. Each truss can be built up to a maximum of three panels in height, three in thickness.

Through the regularity of its openwork pattern and its development in bold horizontals, the Bailey truss lends itself to handsome effects. The trusses pictured above replace wrecked arches of an old Italian bridge. The short top tier, placed where maximum strength is needed in a simple (as opposed to a continuous) beam, gives the bridge an unexpectedly bold and lively shape.

Among military bridges, however, Caesar's pile-and-trestle type was relatively as ingenious, and Xerxes' Hellespont bridge, with hundreds of high-prowed triremes and penteconters serving as pontoons, must have been far more spectacular.

Lift bridge, Japan. c. 1933. 69-foot span.

Counterweights within the tower legs regulate a movable span distinguished by unusual smoothness of shape and surface. A series of rigid frames without diagonals, this is known as a Vierendeel truss in honor of its inventor and chief promulgator, the late Professor Arthur Vierendeel of Belgium.

This Japanese adaptation is shown in preference to any of the hundred-odd examples built since 1896 in Belgium and the Belgian Congo because it best suggests the very considerable esthetic potentialities of the unique construction. The early Belgian Vierendeels were riveted rather than welded, therefore lack the smoothly flowing lines and planes of this all-welded structure; and the more recent Belgian examples, though welded, have a full-curved upper edge that gives them the appearance of arches and is much less forceful than the straight-ended truss shown here. Moreover, the roadway in Belgian practice is normally placed high in the truss, confusing its outlines.

The few Vierendeel trusses that have been built in the United States are so heavily dimensioned as to appear brutal.

Engineers and architects are beginning to explore the new opportunities for structure and shape that are offered by electric arc welding. If the visible effect upon bridge design has thus far been negligible, it is because welding has been used to lighten and smooth familiar bridge forms rather than as basis for the creation of new structural shapes.

By eliminating old-fashioned rivets and the angular joints that accompany them, welding allows a one-piece homogeneous structure with a continuous flow of forces from one part to the next. The relative efficiency of this assembly method is illustrated by the Vierendeel truss pictured above, which weighs a fifth less than a conventional riveted truss of similar design.

Welding implies continuity of structure. When this potentiality is more fully realized we shall have steel bridges unlike any we have known. Steel will be formed into thin sheets, stiffened by bending, and these light shell-like structures will have the strength to span great distances in one smooth leap. They will look more like reinforced concrete bridges of advanced design than like the steel bridges of today, for they will be based on the same principle of structural continuity.

The vertebrate, now supreme, will be challenged by the crustacean.

Double-leaf bascule for the Canadian Pacific Railroad, between Sault Ste. Marie, Ontario, and Sault Ste. Marie, Michigan. 336-foot span.

The great steel arms with their massive counterweights are so nakedly expressive of their capability of sudden movement that the bridge seems like some giant insect.

Pulaski Skyway over the Hackensack and Passaic Rivers, Hudson County, New Jersey. 1932. For the State of New Jersey: Jacob L. Bauer, chief engineer. Main span of 550 feet.

The Skyway undulates high over the Jersey meadows, its continuous trusses swung overhead where extra clearance is needed beneath. Discounting the whimsy of its pier design, it is more plausible than most serpentine trusses for it makes no effort to appear as an arch.

Bridge over the Rur at Düren, Germany. Built 1930; destroyed in World War II. 256-foot span.

This was a bridge without right angles. The triangle of the opening was repeated in the arrangement of truss members and in the design of railing supports. A curiosity anticipated by Brunel in his Chepstow Bridge of 1852, shown on page 81, this triangular truss was recommended by prominent German engineers as economical for medium-length spans.

Firth of Forth Bridge, Scotland. 1883-89. John Fowler and
Benjamin Baker, engineers. Two 1700-foot main spans.

The idea of the cantilever is ancient in the Orient (see page 31), and
the German invention of the modern metal cantilever truss dates back
to 1867, but the fabulous Forth bridge was the first major example.
For thirty years its spans were the longest in the entire world.

Few other bridges approach it in dramatic content. The great
tapered towers with their outstretched cantilever arms have a splendid
sweeping fullness, and their assembly of large tubular members makes
their structure extraordinarily intelligible. Small truss members would
have been confusing in these irregular shapes. Every element of
the design is clearly articulated, from the four separate circular piers
under each tower to the arched "suspended spans" that join the
tips of the cantilevers. Even the difficult juncture with the latticed
side-spans is accomplished without fumbling.

The Forth Bridge is not conventionally pretty or graceful, but there is
a deep emotional satisfaction in its powerful lines.

Viaduct at Viaur, France. Before 1903. Designed by the *Société de Construction des Batignolles.* 722-foot span.

The two great cantilevers meet at the center without the introduction of a "suspended span," making an arched opening.

Twin Falls—Jerome Bridge, Arizona. Before 1927.

Here the "suspended span" is the parallel-edged section at the center.

The thinly etched lines of the trusswork contrast very happily with the massive walls of the canyon.

Grand Glaize River Bridge, Missouri. 1930. Sverdrup & Parcel, consulting engineers.

The fully developed cantilever truss is startling in this underslung version. The piers are now almost entirely submerged.

Cooper River Bridge, Charleston, South Carolina. 1920.
Waddell & Hardesty, engineers. Two separate cantilever trusses,
the larger with a main span of 1050 feet.

This is certainly not the most beautiful bridge in the world. But
perhaps it is the most spectacular, for here is a highway recklessly
launched into the sky. Steep approaches, stupendous height, extremely
narrow width and a sharp curve at the dip conspire to excite and
alarm the motorist, even while his changing perspective of the second

span gives him multiple awareness of the structure that is hurling
him through space. Perhaps all bridges should be bent at the middle
so that no one might traverse them unaware.

From any viewpoint the long unbroken thread of road somehow
manages to tie together the disparate means of support, and the
looming batlike shapes of the cantilever trusses dominate the skyline for
miles around. This is a bridge for the collector of bridges.

Bridge over the North Elbe at Hamburg, Germany. c. 1882.
Three 330-foot spans.

Sagamore Bridge over the Cape Cod Canal at Bourne, Massachusetts. 1935
Fay, Spofford & Thorndike, engineers.
A more graceful version of the ambiguous serpentine motif.

The bridges illustrated on these and the two following pages seem to have no one clearly dominant structural idea: in the same span the beam may, without much show of favoritism, be combined with cable or with an arch, or arch and cable may be used together, or even all three at once. Yet all of them are finally better classified as beams than as anything else, for even the most complicated struggle of forces comes, in the end, to a neutral deadlock, with little or nothing in the way of external push or pull.

Since some of these hybrids are very good-looking, particularly those shown on pages 82 and 83, one cannot say that mixed structure is in itself evil. But when the play of forces is so equivocal that the role of the various members is visually unintelligible, and the different parts seem mutually contradictory, then the design loses conviction and the bridge is more curious than beautiful.

Bridge over the Wye at Chepstow, England. 1852. I. K. Brunel, engineer. Main span of 300 feet with three 100-foot side spans.

Each of the two tracks is separately supported, its girders stiffened in the long jump by bar-chains hung from either end of an iron tube, and provided with light vertical and diagonal stiffeners. The tube resists the pull of the chains, and the final result is a truss with a triangular section.

Royal Albert Viaduct over the Tamar at Saltash, England. 1859. I. K. Brunel, engineer. Two 455-foot spans.

The elliptical iron tubes, 16 feet wide by 9 feet high, act as arches, and the bar-chains absorb their thrust, perhaps also provide support, but it is all very mysterious. With its extreme height and extreme narrowness, accentuated by the slim verticality of the masonry, this Victorian grotesque manages considerable appeal.

Pont Transbordeur, Marseilles, France. 1905. Arnodin, engineer.
787-foot span.

The French flair for boldness and lightness in metal construction is evident in the Transbordeur, which at first sight is a mysterious arrangement of lines in space—without substance or apparent function. Actually it is not a bridge but a support for an aerial ferry that travels from one bank to the other, suspended just a few yards over the harbor waters.

From the high towers hang two cantilever beams, joined by a trussed "suspended span" and anchored firmly to the ground at their far ends. Compare with the cantilevers shown on pages 77 and 79.

Bridge over the Rhine between Cologne and Mülheim, Germany. 1929; destroyed in World War II. Karl Möhringer, engineer. 1033-foot span.

The Mülheim bridge was the finest example of the self-anchored suspension bridge, a type of beam-and-cable construction that is more popular in Europe than the pure suspension bridge. The usual external anchorages are not needed, for the wire cables are attached at either end to the stiff plate girders of the roadway itself. Since these girders must absorb the pull of the cables and also provide much of the actual carrying strength of the bridge, they are rather substantially dimensioned for American taste. But once the structural premises are accepted, it must be admitted that the bridge was very skillfully designed: proportions were excellent and the towers, hinged at the base for flexibility, were unusually clean in outline.

St. Georges Bridge over the Chesapeake & Delaware Canal, Delaware. 1941. Waddell & Hardesty, engineers; Aymar Embury II, consulting architect. 540-foot span.

Here the beam ties together the ends of the arch, removing the need for abutments and thus performing a function similar to that of the beam which anchors the cable ends of the German bridge pictured above. In both cases the roadway girders dominate the design, and the arch or cable is rightly subordinate.

Appearance has been considered with a care not usually lavished on American steel bridges. Note the neat K-bracing of the arch ribs and, even more important, the continuation of the lines of the approach girders in the girders of the main span. This sense of continuity, lacking which a bowstring arch looks clumsy and lifeless, would have been enhanced if it had been possible also to extend the detail of the approach girders—their vertical stiffeners and projecting sidewalks.

Bridge for the Renault Factory, over an arm of the Seine at Billancourt, France. c. 1932. Etablissements Daydé, engineers.

The two suspended cantilever beams, separated by a "suspended span," make a construction readily comparable with that of the Marseilles Transbordeur, but that is the end of any resemblance.

The bridge is beautifully balanced in design, though one might wish that the suave plate girders had been strong enough to do the job by themselves, without the assistance of the bar-chains.

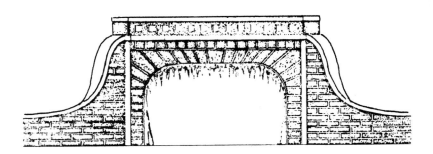

Composite of steel and masonry, reinforced concrete is often treated as a cheap substitute for one or the other of these ingredients. Its own separate character comes out only in the hands of an understanding and sympathetic designer, but then it can emerge in shapes of rare beauty and distinction.

Plain poured concrete, an ancient concoction of cement and water, sand and gravel, hardens in molds to become artificial stone. Like natural stone, it is strong only in compression, therefore suited only to the construction of massive piers and arches. Not until 1875 or so was it discovered that this man-made masonry might be given strength in tension through the incorporation of embedded rods of iron or steel. Thus was born a new and scientific material, reinforced concrete, with the compressive strength of stone, the tensile strength of steel, plus a plastic quality that is entirely its own and most appropriately expressed in a fluid continuity of structure and line.

Steel-reinforced concrete is a patient material, all too tolerant of torture. The plasticity that is its great advantage is also a weakness, for it permits all kinds of gross indignities. Illustrated here are some of the ways in which it has been abused through the years; the imitations of stone construction are as patently absurd as the built-in stalactites shown above; the clumsy truss suggests that structural forms well suited to assembly from lengths of timber or steel may be foreign to reinforced concrete, and the elaborate foolishness of the pseudo-modern Connecticut underpass is obvious as such. Compare these dismal structures with the beautiful bridges on the two following pages, both completed in 1905, proof that even at that early date there were two engineers—Francois Hennebique of France and Robert Maillart of Switzerland—who were successful in creating structural shapes eloquent of the unique powers and properties of the wonderful new material.

Good spare construction is not easy in reinforced concrete. At every stage it requires a great deal of skill, care and sensitivity—from the workmen who make the forms, place the reinforcing steel and mix the concrete as well as from those responsible for design and supervision. Small wonder then that the best work has been done in Europe, where thrifty use of material has long been essential and where loving craftsmanship is still something of a live tradition.

MISUSE OF REINFORCED CONCRETE:

Alvord Lake Bridge, Golden Gate Park, San Francisco. 1889. 20-foot span.
This first reinforced concrete bridge in the United States is still standing. It is less remarkable for its imitation of rusticated stonework than for its custom-made stalactites.

Arlington Memorial Bridge, Washington, D. C. 1932. John L. Nagle, engineer; McKim, Meade & White, architects.
The bridge is designed in Washington's usual pompous neo-classic manner. Its open-spandrel arches of reinforced concrete are faced with granite slabs in faithful imitation of solid stone vaults, and its central draw-span of steel is painted and decorated in faithful imitation of the aforesaid faithful imitations.

Ridge Road Bridge, Wethersfield, Connecticut. c. 1938. By the Connecticut State Highway Department.
This imitation in reinforced concrete of a medieval stone bridge, such as that illustrated on page 17, is as inept as it is absurd. Note the use of pointed cutwaters to divide the traffic lanes.

Reinforced concrete is molded to form a lengthy truss, gross indeed as compared with the ordinary steel truss visible at the far left of the picture.

Merritt Parkway Underpass at Stamford, Connecticut. c. 1937. By the Connecticut State Highway Department.
This rigid frame of reinforced concrete apes no historical precedent. Its vulgar ornament is peculiar to our times and easy of achievement in this docile material.

Bridge over the Ourthe, Liège, Belgium. Built during four winter months for the
Liège Exposition of 1905. François Hennebique, engineer. 180-foot span.

Hennebique (1842-1921), a French engineer-contractor celebrated for his early
development of reinforced concrete construction, was one of the first to realize that the
new material lent itself to a smooth flow of structure and surface.

In its lean elegance this bridge has had few rivals. The flattened arch becomes
amazingly thin at the crown, yet this photograph taken on the official proving day
shows that it was capable of supporting three steam-rollers.

Tavanasa Bridge over the Rhine, Canton Grisons, Switzerland. 1905; destroyed by landslide in 1927. Robert Maillart, engineer. Three-hinged arch of 167-foot span.

Following his master, Hennebique, in the quest for integrated structure, Maillart fused arches and road slab to form a structural unit, proudly revealed. Further explanation of the construction will be found on page 106.

This was Maillart's first masterpiece.

Detroit-Rocky River Bridge, near Cleveland, Ohio. 1911. 280-foot span.
Arch ribs are plain concrete, but the rest of the bridge is reinforced.
Like the other first large American arches in the new material, this
was closely patterned after a famous stone bridge of 1903:
the Pont Adolphe at Luxembourg, a twin-ribbed open-spandrel arch
of record 280-foot span. The more appropriate models of Hennebique
and Maillart (pages 86 and 87) were disregarded then as now.

Bixby Creek Bridge on the Carmel-San Simeon Highway, California.
1933. By the California Division of Highways. 320-foot span.
The great arch ribs are dwarfed to insignificance by colossal
abutment piers, a misplaced emphasis that distorts the balance of the
bridge and destroys its continuity of line. Compare with the more
recent bridge by the same office that is shown on the facing page.
Thickened piers over arch abutments are generally rationalized as
wind bracing, but since they have been proved dispensable their
continued popularity in this country seems attributable to nostalgia for
the monumental forms of ancient stone construction.

It is in arches that reinforced concrete achieves its boldest spans. Occasionally the roadway is suspended from the arch. More often it runs above, as in the bridges shown on the next few pages. Since plain concrete masonry, like stone or brick, is strong in compression, little or no reinforcement is required for a massive fixed-end arch, but only skillfully embedded steel makes possible limber two and three-hinged arches, slender spandrel columns or cross-walls, and thin flat decks.

Typical of today's best standard practice are the graceful arches shown on the following pages. They are so undeniably handsome that it is perhaps ungrateful to complain of their somewhat indifferent relationship to their material. Assembly of apparently separate pieces—supporting ribs, intermediate posts and supported, seemingly inert deck—conveys little feeling of the unity and continuity of structure and of shape that is implicit in reinforced concrete, and represents an advanced stage of the development toward lighter and more economical masonry construction that started with the medieval Céret arch (page 24) rather than design freshly conceived in the specific terms of a totally new material. A typical offender in this special sense is the otherwise admirable Russian Gulch Bridge (opposite). Compare it with the work of the independent master, Maillart (pages 87 and 102-113), where steel and masonry, arch and superstructure, are so completely fused into a single working shape that execution in any material other than reinforced concrete is unthinkable.

Russian Gulch Bridge on the Mendocino Coast Road south of Fort Bragg, California. 1940. By the California Division of Highways: F. W. Panhorst, bridge engineer. 240-foot span.

The valley is spanned in one graceful gesture, for the posts that support the roadway march without break over banks and arch, waning in size as they approach the crown and waxing as they take the downward path. The complete separation of arch from roadway is very pleasing in this high-reaching elliptical arch, whereas flatter arches look best when they are joined with the roadway at the crown.

We are so accustomed to considering good spare concretework as a prohibitively expensive luxury in the United States that it is gratifying to hear that this beautiful bridge was judged the most economical solution to the problem.

Sandöbron (Sando Bridge) over the Angerman River, Sweden. Built 1937-42. By the Skånska Cement Company: S. Håggbom, chief engineer. 866-foot span.

This is the longest reinforced concrete arch in the world. Pairs of round columns support the long approach viaducts (see also page 116) and lift the roadway over the mammoth single-ribbed arch.

Traneberg Bridge, Stockholm. See opposite page.

Traneberg Bridge, Stockholm. 1934. For the Stockholm Harbor Board:
Ernst Nilsson and S. Kasarnowsky, engineers; Eugène Freyssinet,
consultant; Paul Hedquist and D. Dahl, architects. 585-foot span.

The transverse slab-walls that carry the deck obstruct the diagonal
view, giving an illusion of mass, yet they seem more appropriate to
reinforced concrete than the isolated posts of the Sando Bridge on the
opposite page, and remove the otherwise startling resemblance to

steel construction. These cross-walls carry the roadway smoothly over the
twin-ribbed arch and beyond, interrupted only by the fusion of deck
and arch at the relatively low crown. There is no special monumental
treatment at the abutments.

Note the crisp molding about the upper edge of the arch ribs, also
the cleanly cantilevered sidewalk with its light steel railings.

Kungsbron, Stockholm.
A close view of one of the flat-arched twin-ribbed spans of the unusual double bridge that is shown on the opposite page.

Royal Tweed Bridge, Berwick, England. 1928. L. C. Mouchel and Partners, engineers.
The rhythmic continuity of the long flat arches and the spandrel posts is unbroken by any special accent over the abutment piers. Note that the span of the arches increases as the roadway mounts from the low bank on the right to the higher bank on the left.
The juncture of arch crown and deck, never an easy problem, seems somewhat tentative as compared with the Swedish bridges, but the treatment of spandrel walls and posts as a smooth continuous plane, recessed behind deck and arches, has considerable merit.

Kungsbron (King's Bridge), Stockholm. c. 1940. For the Stockholm Harbor Board:
A. Wickert & S. Kasarnowsky, engineers.

    Since a broad-decked short-span bridge is bound to look stubby and awkward, this
small urban bridge was molded in two completely separate sections, each carrying
one-way traffic. The longitudinal split also allows the bridge to fit easily into its
man-scaled surroundings.

93 REINFORCED CONCRETE ARCH

Gueuroz Bridge over the Trient Glacier, Canton Valais, Switzerland.
1933. A. Sarrasin, engineer. 323-foot span.

Reinforced concrete has been cast into a working shape of
extraordinary visual power, closer in spirit to the work of the older Swiss
engineer, Maillart (pages 102-113), than to the other arches shown
in this section.

Parapets are usually inert, extraneous elements, but here they serve
as beams to support the approaches and to stiffen the slender arch
ribs. This interdependence of structural members is expressed in the
smooth flat plane formed by the parapet, posts and arch.

The refinement of line and proportion is all-pervasive. Note the
vigorous shape of the voids formed by the rounded juncture of posts and
parapet, and the fine relationship of the spandrel posts to the sturdier,
more widely spaced verticals that carry the approach spans.

Gueuroz Bridge, Switzerland.

A worm's-eye view of the distinguished bridge that is illustrated on the opposite page.

The spacing of the braces that tie together the two arch ribs has been handled with exceptional neatness.

Proposed bridge over the Rhône at St. Maurice, Switzerland. c. 1945. A. Sarrasin, engineer. 328-foot span.

The amazing thinness of the deck is due to its unusual construction as an active self-supporting slab of reinforced concrete, growing out of the rectangular mushroom-headed columns that carry it over arch ribs and river banks. There are no beams.

This type of mushroom-slab construction, with its smooth flow of line from column into slab, was invented by Maillart around 1908 (see page 102) and used by him in buildings of many kinds—never, however, in his bridges. Maillart preferred to reveal his thin slabs as such only in the substructure, and thus kept the apparent weight of his bridges high, with gratifying results. His stiffened slab-arches (pages 102-104), quite the reverse of this project in principle, have nothing of its droopiness. Their emphasis is upon the firm continuous line of the roadway itself, not upon the means of its support.

Another notable feature of this design is the use of paired columns for wind-bracing at the abutments instead of the customary and ungainly thickened piers.

The United States has a bridge of similar construction—the Fort Snelling-Mendota Bridge built over the Minnesota River in 1926, designed by C. A. P. Turner and Walter H. Wheeler. But it is far less graceful than this Swiss project, partly because it uses Turner's system of mushroom-slab construction (patented in 1905) in which columns are separated from the slab by rectangular plinths or capitals.

La Roche-Guyon Bridge over the Seine, France. 1934; destroyed in World War II. Etablissements Boussiron, engineers. 528-foot span, a record for overhead reinforced concrete arches.

Even in a country with a long tradition of fine reinforced concrete construction this arch was of outstanding merit.

Arch ribs are of smoothly varied section: at their spring they are flat hollow rectangles; at road level they are solid and square; at the crown they again become hollow rectangles, but now emphatically vertical. The crescent shape, unusual in hingeless arches, is particularly effective in the slow-rising curve used here, and the light lattice-bracing avoids any feeling of top-heaviness.

Note that the hangers (pure tension members) are of reinforced concrete and designed to match the posts (compression members) that support the approaches. This note of formalism should probably be condemned, yet, as in the Chicago overpass pictured on page 53, it contributes a great deal to the unity and harmony of the design. Light steel hangers would have been more reasonable, but in that case the visual center of gravity would have slipped below the roadway and upset the balance of values.

Bridge over the Kalix älv, northern Sweden. 1933. Skånska Cement Company, engineers.
   The overhead arch as an impressive demonstration of the power of masonry.

The emotional impact of the overhead arch is unique. The more sharply defined and proudly isolated, the more monumental its effect. It is not so much an assertion of reinforced concrete, for steel plays a very minor role in its compressed curve, as it is the ultimate demonstration of the brute power of plain masonry. The United States offers nothing comparable in excellence with the European examples illustrated here, for the American designer, instead of playing up the lightness of hangers and roadway and railings as legitimate contrast to the massive supporting arches, tends to extend the weightiness of his arch ribs to each smallest detail and thereby kills the spirit of his structure.

Bridges with overhead structural members—suspension bridges excepted—are a difficult problem in any material, for there is always the danger that vital cross-braces will seem either distractingly complicated or oppressively bulky. Compare with the steel overhead arches and trusses shown on pages 50 to 53.

Bridge over the Seine at St. Pierre du Vauvray, France. 1922; destroyed in World War II. Eugène Freyssinet, engineer. 430-foot span, then a record for reinforced concrete.

To pass over this famous bridge must have been a startling experience, for the roadway was scarcely wider than the sum of the two arch widths. One would first have felt squeezed between the giant ribs, then liberated as they soared up through space, free of cross-bracing other than the heavy frames at either portal.

Suspended from the hollow cellular arch ribs were the wire hangers, thinly coated with cement mortar, that supported the light trusses that in turn carried the roadway. The filigree of the trusses was recalled in the design of the railings.

Sörängsbron (Sorangs Bridge), Hälsingland, Sweden. 1930. Skanska Cement Company, engineers. Three-hinged arch of 177-foot span.

Inclined hangers, a Scandinavian invention, reduce the stresses in the arch ribs and permit their extreme slenderness.

The three-hinged arch was cast as two entirely separate pieces. The exceptional neatness and lightness of arches, hangers and deck make the awkward design of the abutments seem all the more unfortunate by contrast.

Proposal for a thousand-meter (3281-foot) arch of reinforced concrete.
Eugène Freyssinet, engineer.

The substantial reputation of its designer lends authority to this daring project.
In 1928 Freyssinet proposed such a bridge for the Hudson River.

Divided, splayed haunches contribute to the grace of the tapered curves, and the
level at which the roadway and arches intersect seems very well chosen.

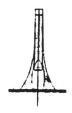

Prestressed concrete bridge of Luzancy.
page..
The hollow rib that is being hoiste.... .... p.... ...
this picture is made up of many short precast concrete
sections, compressed by taut steel wires to take up
dead load. No scaffolding is necessary. When the rib
is in position longitudinal wires will be threaded through
from one abutment to the other and tightened to pull
the sections into compression and form an arch.
The total prestressing will allow for moving loads as well
as for the weight of the bridge itself.

Supports for a third rib are evident at extreme left
and extreme right. The thrust is transmitted to the
massive abutments through the split arch ends. Because
of this articulation at the abutments the bridge is
called a *pont à béquilles*, or "crutch bridge."

As sometimes happens in technically advanced
reinforced concrete construction, the various elements
of the bridge are interwoven in such complex fashion
that its exact structural nature is somewhat controversial.
It may be termed an arch, but from another point
of view it might better be described as a rigid frame
of very unusual type.

Prestressed concrete bridge at Luzancy.
An advanced stage of construction. The upper
and lower surfaces of the three hollow ribs have
been joined with precast concrete slabs and
transverse wires.
The completed bridge is illustrated on the
opposite page.

Prestressed concrete bridge at Luzancy, Seine and
Marne, France. 1946. Eugène Freyssinet, engineer.

The leadership of France in reinforced concrete construction has been due pre-
eminently to Hennebique (page 86), to Auguste Perret, the architect, and to Eugène
Freyssinet, celebrated for his vaulted hangars of 1924 at Orly, his great arched bridges,
one of which is shown on page 99, and now for his work with *prestressed concrete*.

The prestressing of concrete consists in artificially creating stresses approximately
equal and opposite to those that are produced by dead weight and live load in the
completed, functioning structure. In the Freyssinet system this is effected through the
pressure exerted by a series of parallel steel wires of high tensile strength that are
stretched to the limit of their elasticity and embedded in the concrete, thus creating
permanent compression. Such construction uses at least 70% less steel than ordinary
reinforced concrete, an economy of financial importance; and it uses 30 to 40% less
concrete, an economy of esthetic importance, for it implies unprecedented slenderness.

Freyssinet advocates the use of prestressed concrete in prefabricated sections.
Entire factory-made beams are feasible for small and medium spans. Long beams or
arch ribs, however, are assembled on the site from precast concrete units of easily
handled size. These are placed end to end and joined by wire threaded through holes
provided for the purpose. The joints between units are then filled with mortar and the
cables stretched taut. When the beams or arches, whether one-piece or composite, are
all in place, they are threaded together transversely with more stretched wires.

The system found its first major proving ground in wartime Tunisia, where June, 1943,
found the reinstated French with 300 bridges that needed immediate repair or replace-
ment, and with very little steel and almost no wood for forms. Prestressed concrete in
prefabricated units was the logical answer and proved successful far beyond its
emergency value. The Tunisian bridges were not particularly attractive, but the visual
possibilities of the construction are suggested by the handsome new bridge at Luzancy
that is pictured here.

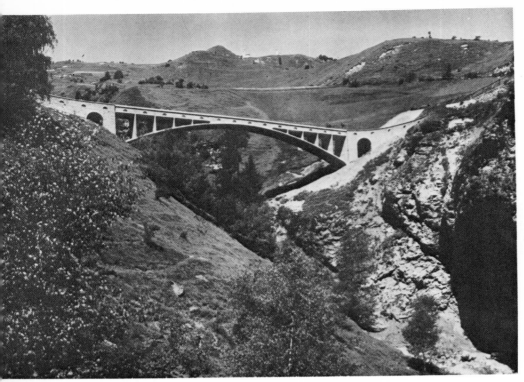

Val Tschiel Bridge, Canton Grisons, Switzerland. 1925-26. Robert Maillart, engineer. Stiffened slab-arch of 142-foot span.

Here for the first time Maillart used his extremely thin barrel vault, stiffened by the deep rigid girder formed by parapets and road-slab.

The semicircular openings along the parapet and the incongruous masonry abutments were stipulated by the Grisons officials.

Robert Maillart (1872-1940) used reinforced concrete to enter wholly new realms of structure and shape. This Swiss engineer was so far in advance of his time that the full meaning and impact of his work may not be felt for years to come. Whereas most engineers tend to be enslaved by their formulas, Maillart used his formidable scientific and technical knowledge as the tool of his intuition, creating structure that transcended accepted patterns and limitations to reveal the laws of nature in terms of new and surpassing beauty. It is not surprising that the forms at which he arrived with notable independence should be kin to those evolved by other great modern artists working in other and very different media.*

Early experience under François Hennebique, French engineer famous for his pioneer work with reinforced concrete (page 86), must have sharpened Maillart's own awareness of the plastic character of the new material and its irrelevance to structural shapes traditional in stone or metal. Out of his profound understanding of reinforced concrete he gradually developed new and magnificently appropriate types of construction in which each small part became an actively participating member of an organic whole.

Although Maillart designed many buildings, his greatest achievement was his bridges. Most of them are small and hidden in remote Alpine valleys, for his work was too unconventional to receive much support; but his spans were relatively so inexpensive —because of their efficient use of material—that officials could not afford to ignore him entirely. These bridges are in the main of two distinct types: the *stiffened slab-arch*, illustrated here, and the *three-hinged arch with integrated road-slab*, shown on pages 106 to 111. His great invention of the *mushroom slab*** he did not himself apply to bridges, but a proposed application by a younger engineer is illustrated on page 95.

*A stimulating study of these relationships, outside the scope of this book, may be found in Space, Time and Architecture, by Dr. Sigfried Giedion.

**The most complete presentation of Maillart's work will be found in a monograph by Max Bill that has very recently been published in Switzerland.

Maillart's *stiffened slab-arch* is something like a reversed suspension bridge in its structural action, for the flexible vault, of eggshell thinness, takes only direct thrust, while road-slab and parapets together form a rigid U-shaped girder that resists local bending under concentrated moving loads. Spandrel supports, too, are thin continuous slabs, active in all three dimensions. Characteristic of Maillart's use of reinforced concrete is this emphatic insistence upon the slab as the basic element of construction, far more appropriate to the material than the usual steel-inspired network of isolated posts and isolated beams.

The miraculous lightness of these bridges must be attributed to the extraordinary efficiency of their revolutionary construction, their equally miraculous elegance, to the consummate artistry of their designer.

Footbridge near Wülflingen, Canton Zurich, Switzerland. 1933. Robert Maillart and W. Pfeiffer, engineers. Stiffened slab-arch of 124-foot span.

The subtle reverse curve of the footway makes this one of the most graceful of Maillart's bridges. Arch and deck slabs fuse into one as they approach mid-stream, yet the total thickness at the crown is only 4½ inches.

Schwandbach Bridge. See opposite page.
   The inner edge of the slab-arch follows the curve of the highway.

Schwandbach Bridge.
   The outer edge of the arch, straight in plan, serves as base for the sloping sides of the cross-walls that support and brace the curved deck girder.

Schwandbach Bridge, near Schwarzenberg, Canton Berne, Switzerland. 1933.
Robert Maillart, engineer. Stiffened slab-arch of 111-foot span.

A curiosity among bridges, this curved span is a dramatic example of Maillart's extraordinary feats of engineering. As demonstrated by the photographs on the preceding page, the inner edge of the vault follows the elliptical curve of the deck, but the outer edge is straight, with vertical cross-walls brought up on a diagonal to buttress the bridge against centrifugal action.

The thinness of the arch (7.9 inches) and the cross-walls (6.3 inches) looks precarious to anyone unfamiliar with the fantastic strength of reinforced concrete in favorable construction. A special advantage of the extremely light vault of these bridges is the minimum of expensive scaffolding that is required for support during construction.

The scaffolding of the Salgina Bridge. See opposite page.
    The daring lightness and notable elegance of the arch were presaged
in its scaffolding, designed and executed by the Coray family of Chur, long
famous in this highly specialized field.

Maillart's other great bridge type was the *three-hinged arch with integrated road slab,* suited through its elasticity to greater spans and to less stable foundation conditions than the stiffened slab-arch illustrated on the preceding pages.

Joints at mid-span and at either abutment divide a three-hinged arch into two symmetrical halves, each of which should be thickest at its center if it is to make most efficient provision for moving loads. With its bulging ribs under a separate, passive roadway, the usual arch of this type is singularly ungainly. Even while he reduced its weight and heightened its effectiveness, Maillart transformed the three-hinged arch into a thing of beauty. As early as 1905, in his Tavanasa Bridge (page 87), he used a reinforced concrete road slab as an active structural member, fusing it with his open U-shaped arch ribs at the critical quarter points to form a strong, closed box-shaped girder that tapered to the hinge at the crown.

He developed innumerable variations upon this theme. The solid masonry abutment piers of the Tavanasa arch were replaced by thin cross-walls of reinforced concrete, similar to those he used as supports between arch ribs and deck girder. In his latest bridges these transverse slab-walls, and the arch ribs too, assumed vigorously curved and angled outlines as Maillart shaped them to extract the utmost strength and meaning from his material.

The significance of these bridges goes deeper than the lithe elegance of their appearance or the technical virtuosity of their structure, for in them, by grace of their creator, reinforced concrete is quickened to life and given a voice unmistakably its own.

Salgina Bridge, near Schiers, Canton Grisons, Switzerland. 1930.
Robert Maillart, engineer. Three-hinged arch of 269-foot span.
    A classic version of its structural type, the lean and flattened arch
bridges the chasm in one smooth leap.

107 **REINFORCED CONCRETE ARCH**

Bridge over the Aare at Innertkirchen, Canton Berne, Switzerland. 1934. Robert Maillart, engineer. Three-hinged arch of 96-foot span.

This small bridge is of different construction and rather gentler demeanor than Maillart's customary three-hinged arches. His usual U-shape is reversed here, for spandrel walls and road slab together form a girder that is open beneath.

Bridge over the Thur, near Felsegg, Canton St. Gall, Switzerland. 1933. Robert Maillart, engineer. Two parallel three-hinged arches of 226-foot span.

No bridge of Maillart's is more assertive of strength than this light span over the Thur.

Characteristic of his later work are the powerful, concisely defined, highly differentiated shapes—the pointed arch, the straight-drawn outer edges of the arch ribs, the splayed slab-supports of the approaches. Every part is alive and at work.

The X-shaped abutment-joints of reinforced concrete, more economical than conventional steel hinges, contribute a great deal to the sense of unity and continuity of structure.

Bridge over the Thur. See also opposite page.
    At once bold and delicate, the lithe arch looks very much at ease in
the friendly, man-scaled landscape.

109 REINFORCED CONCRETE ARCH

Bridge over the Arve, near Geneva, Switzerland. 1937. Robert Maillart, engineer. Three parallel three-hinged arches of 194-foot span.

In paring the substance of his bridges to an irreducible minimum, Maillart evolved extraordinary new forms. Like the abutment hinges that they so much resemble, these X-shaped road supports combine elasticity and strength with economical use of material.

The light railing, assembled of rolled steel sections, is a model of propriety.

Lachen Bridge, Altendorf, Canton Zurich, Switzerland. 1940. Robert Maillart, engineer. A skew bridge with two separate off-set three-hinged arches.

Arches spring from different levels to carry a highway over railroad tracks at a sharp angle, and the static symmetry that we take for granted in a bridge span is replaced by a dynamic interplay of shapes.

Bridge over the Simme, Garstatt, Canton Berne, Switzerland. 1939. Robert Maillart, engineer. Three-hinged arch.

The smooth curves of the Salgina arch (page 107) hardened nine years later into these taut diagonals—strong and decisive, infinitely expressive.

The gabled roof in the background belongs to an old wooden bridge.

REINFORCED CONCRETE ARCH 110

Bridge over the Arve. See also opposite page.
    The very slightly curved arch ribs meet in a point at the crown and the bridge becomes wholly expressive of its tri-jointed construction.

Gündlischwand Bridge, Canton Berne, Switzerland. 1937. Robert Maillart, engineer.
A skewed continuous beam with main span of 125 feet.

    Maillart's bridges were not invariably arches, but they were always of reinforced
concrete; and whatever their structural principle they were imbued with their
designer's acute awareness of the unique nature of his chosen material.

Châtelard Aqueduct, Canton Valais, Switzerland. 1925. Robert Maillart, engineer.
100-foot span.

    The structure is hybrid, for the arch springs conventionally from its abutments, then merges with the box girder that carries the water.

    This fusion of two seemingly incompatible forms is curiously successful.

113 REINFORCED CONCRETE: BEAM AND RIGID FRAME

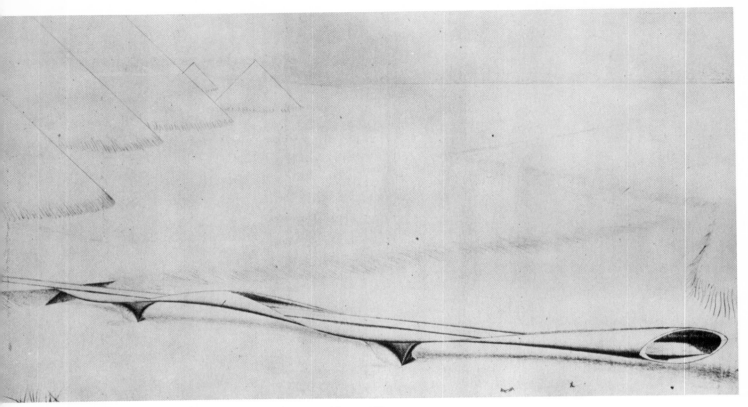

Proposal for a long-span highway bridge of reinforced concrete. 1948.
Paolo Soleri, architect. Continuous beam.

The undulated slab flies over the river like some strange sleek bird. Its winged flanges are convex at the piers, then soar up and over in a reverse curve to embrace the roadway at midspan. There are no separate elements—only the attenuated multi-curved slab, one with the piers from which it springs.

Essentially the bridge is a tube—carved away where superfluous and turned inside out at the piers.

Maillart showed how a reinforced concrete bridge might become one thing, how it might grow out of the fluid, continuous character of its material. It was in this spirit that he developed his principle of the slab. . . . Perhaps it had to be an architect, committed to the creation of space by the very nature of his art, who would take the next step and free the slab to come alive in three full dimensions.

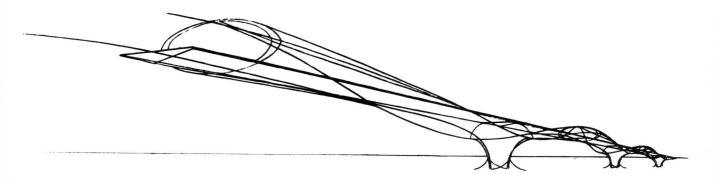

Proposal for a long-span highway bridge. See also opposite page.
    Sketched above is a lineal analysis.
    Reproduced below are elevation, plan, longitudinal section and, at bottom
right, the transverse sections at mid-span, quarter-span and mid-pier.

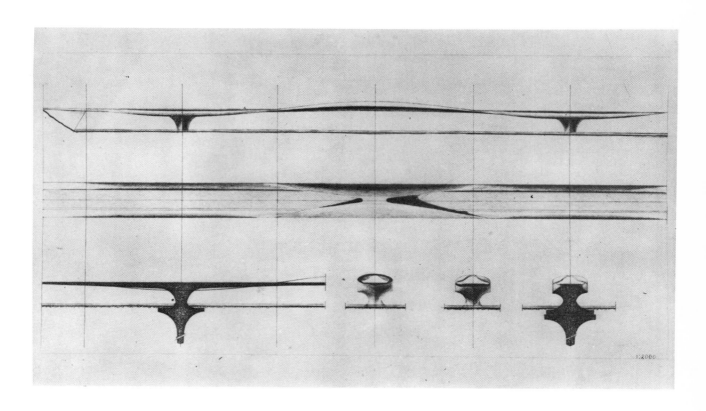

115 REINFORCED CONCRETE: BEAM AND RIGID FRAME

Bull Run Creek Bridge for the Norris Freeway, near Knoxville, Tennessee. 1934. By the Tennessee Valley Authority. Continuous beam with main span of 50 feet.

Through the flat planes of piers and beams the bridge becomes a geometric abstraction of its structural idea.

Approach to the Sando Bridge, Sweden. (See page 90.)

The curving viaduct is so high, so light, so cleanly drawn that its unobtrusive presence actually enhances the quality of the natural landscape.

Bridge over Henderson Bay, Pierce County, Washington. 1937. By Pierce County: F. A. Easterday, engineer. Cantilever beam with main span of 190 feet.

The road slab forms the top of a hollow box girder, and the girder in turn is monolithic with its supports. Note the "suspended span" inserted at center.

The designer has stated his principle of construction in unusually agreeable terms.

Dry Creek Bridge, Wabunsee County, Kansas. 1941. By the Kansas State Highway Department: E. S. Elcock, designer; G. W. Lamb, bridge engineer. Rigid frame with spans of 50, 70 and 50 feet.

The substructure, with its tapered, divided piers, is shaped with exquisite skill, but the design as a whole suffers from the overcomplication of coping and parapet.

Side spans run straight to the abutments, with no semblance of arch construction.

Overpass at Oelde, Germany. c. 1938. For the *Autobahn:*
Karl Schaechterle and Fritz Leonhardt, chief engineers; Paul Bonatz,
architect. 108-foot span.

This simple beam is comparable to the Nazi plate girder bridges
(pages 65 and 66) in its sober refinement and in such specific details as
the shallow, decisively projected sidewalk slab, continued over the
retaining walls as a coping, and the light railing without terminal
accents.

The ends of the beam are not concealed. They rest in full sight upon
their supporting piers, contributing to the notable clarity of statement.

The precast beam looks rather heavy, especially considering the
fact that it was prestressed according to a variation of the Freyssinet
system described on page 101; but then the Germans have never had
the light touch with reinforced concrete that has been so characteristic
of French work.

It is with some justice that flat-spanned overpasses of this type have
been criticized as a psychological obstruction to fast traffic. Arched
openings usually seem higher and safer to a speeding motorist.

Straight beams are reasonable and economical in reinforced concrete, though
inexpressive of its special character. Design problems are much the same as in similar
construction of steel, but elegance is more difficult to attain, for reinforced concrete is
relatively bulky and lacks the vertical stiffeners and the sharply profiled edges that
give delicacy and scale to a steel plate girder. But brutality can be avoided through
precise statement of the structural principle, and through the welcoming of every oppor-
tunity to introduce lightness and fineness as contrast to the dominant mass.

Rarely at complete ease in rectilinear forms, reinforced concrete comes into its own
in continuous beams with curved under-edges (see the discussion of such construction
on page 10) and in rigid frames, where the fusion of verticals and horizontal is particularly well suited to the fluid quality of the material. In either event the construction
assumes its own logical form, which it can do very handsomely, without trying to duplicate the appearance of arches.

Gardiol Bridge above Montreux, Switzerland. 1944. E. Gardiol, engineer. Continuous beam with 48-foot spans.

Like many mountain bridges, this narrow railway viaduct is curved in plan. Its continuous beam construction is of an unusual type, for the beam is not set on rollers, but cast as one with the slim splayed piers, some of which are 98 feet tall. These flexible supports provide the elasticity needed to allow the beam to move in response to temperature changes.

Waterloo Bridge, London. See also opposite page.

Waterloo Bridge over the Thames, London. Built 1939-45 to replace John Rennie's famous Waterloo Bridge of 1817. Rendel, Palmer & Tritton, engineers, in association with Sir Peirson Frank; Sir Giles Gilbert Scott, architect. Continuous beam with five 240-foot spans.

Long leaping curves are executed with such easy grace that the great new bridge, far from disfiguring the ancient face of London, brings it new life, new and exciting perspectives. The dome of St. Paul's dominates the skyline at the right. Here is ample proof that distinguished twentieth-century architecture can take its place proudly in any setting. Compare with the manner in which the designers of our Arlington Bridge in Washington (page 85) solved their somewhat similar problem.

Neither the reeded coping nor the angular motif at the junction of the beam and piers is completely convincing, but the latter may be partially accounted for by the unusual construction: the continuous beams are not set on rollers to allow for movement, but fused with flexible bearing walls that are set within rigid shell-like piers.

The bridge is faced with slabs of Portland stone, laid in vertical courses to avoid any resemblance to solid masonry.

Proposed highway bridge over the Wisconsin River near Spring Green, Wisconsin.
1947. Frank Lloyd Wright, architect. Cantilever beam adaptable to spans up to
200 feet.

The architect calls it a "butterfly" bridge because its outstretched wings
concentrate the load upon a deep central girder. The elegance of the design is
best evident in the cross section at mid-span that is illustrated above: at this point
the substance of the bridge is reduced to the shallow V-shaped structure shown in
solid black, while the shaded portions indicate the increasing depth of the span
as it curves back to the inset piers from which it springs. The heavy longitudinal
girder at the center projects above the deck to separate the traffic lanes.

The outer shell is no inert surface, but a "stressed skin" that works as one with
the light stiffening ribs. Structure becomes continuous, flowering out of the plastic
nature of the material, consistent with Frank Lloyd Wright's conception of
architecture as organic.

The cantilever principle is used lengthwise as well as crosswise. The drawings
on the facing page show that the bridge is conceived as a series of standardized
self-supporting units, each cantilevered out from its central pier to meet the
arms of adjacent units at mid-span.

Most engineers' bridges simply cut through space. Their interest is in flat elevation
rather than in the depth plane. This architect's bridge is quite a different matter,
for it gives space shape and meaning.

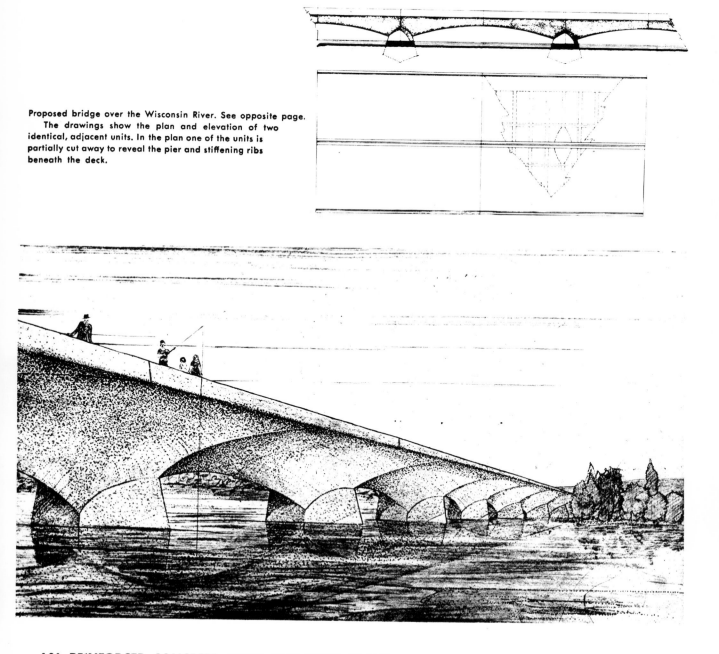

Proposed bridge over the Wisconsin River. See opposite page. The drawings show the plan and elevation of two identical, adjacent units. In the plan one of the units is partially cut away to reveal the pier and stiffening ribs beneath the deck.

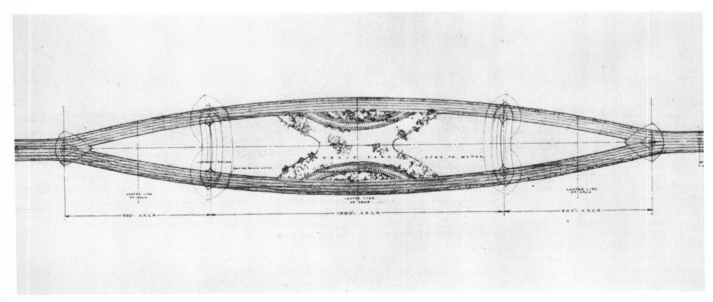

Proposal for a Butterfly-wing Bridge over San Francisco Bay.
    This plan shows the division of traffic into two separate highways as the bridge makes its great triple jump over the main channel of the Bay.
    The two arcs are joined at the center, 175 feet above water, by gardens and parking space disposed on a platform of reinforced concrete.

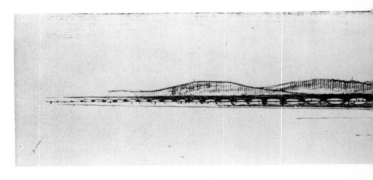

Proposal for a Butterfly-wing Bridge over San Francisco Bay.
    Great shell-like cantilever arms span the three broad openings making a record-breaking thousand-foot leap at the center.

Proposal for a Butterfly-wing Bridge over San Francisco Bay from San Francisco to Alameda, California. 1949. Frank Lloyd Wright, architect; J. J. Polivka, engineer. Cantilever beam with typical span of 156 feet; main spans of 500 and 1,000 feet.

The butterfly-wing principle that Frank Lloyd Wright first conceived for the modest requirements of the Wisconsin River (see the two preceding pages) comes into full flower in this proposal for a southern crossing over San Francisco Bay. The site is the one favored by most authorities as the best means of easing congestion on the existing Bay Bridge.

The body of the bridge is composed of a single repeated cantilever unit similar to that of the Wisconsin project. The roadway is balanced upon a central longitudinal girder that grows out of the tap-root piles. Its spread is reinforced from beneath by thin shells of concrete-sprayed steel mesh that curve up and out from the lower edge of the deep spinal girder.

The soaring double arc of the roadway as it splits and swells outward and upward over the main channel is a brilliant variation and expansion of the established, oft-repeated theme. It is not a true arch, nor does it simulate one. Instead, it is formed by the out-reaching of cantilever arms from the two great U-shaped piers to meet at mid-span, 175 feet above the water. Here the two lightly flying, out-curved halves of the roadway are joined and braced by a landscaped park, fabulous hanging gardens for the delight of citizen and sightseer.

The bridge would be extremely economical in construction and maintenance. Erection of the spans as stiffly reinforced arms obviates much of the costly erection work usual in reinforced bridges, for the stiff reinforcement itself serves as scaffold and centering. The small standardized spans are well adapted to prefabrication, and what little formwork is needed might be used again and again. There would be little maintenance, for there would be no exposed steel to paint and replace. A further advantage claimed by the designers is relative earthquake safety—a claim worth attention when made by the architect of the Imperial Hotel in Tokyo.

The longest span now achieved in reinforced concrete is the 866 feet of the Sando arch in Sweden (page 90). If the people of the Bay region have the foresight to translate vision into reality, they will have a bridge second to none in the world in beauty and in boldness.

GLOSSARY

abutment: An end pier of a bridge, particularly of an arched bridge.

arch: See Structural Types, page 11; also for *fixed arch, two-hinged arch* and *three-hinged arch*.

    segmental arch: An arch curve that forms part of a circle.

    elliptical arch: A curve determined by two foci.

bascule bridge: A drawbridge working on a horizontal pivot.

beam: See Structural Types, page 10; also for *continuous beam*.

caisson: A box or chamber used for construction under water.

cantilever: See Structural Types, page 10.

coping: The capping or covering of a wall.

corbel: A projection from the face of a wall, supporting a weight.

cornice: Moldings run along the top of a wall.

crown: The apex or summit of an arch.

cutwater: An angular or curved structure projecting from a pier that cleaves the water and so lessens its pressure.

girder: A supporting horizontal beam.

    plate girder: A solid-walled metal girder.

laminate: To build up of separate laminae or layers.

lintel: A horizontal beam supporting an opening.

modillion: An ornamental block or bracket under a projecting cornice.

monolithic: One-piece structure: material or materials so brought together as to become an indissoluble structural unit.

parapet: A low wall or protecting railing.

pier: One of the vertical supports of a bridge.

rigid frame: See Structural Types, page 10.

spandrel: The walls between supporting vault and bridge-deck.

truss: Separate members (such as beams, bars or rods) assembled to form a rigid framework.

vault: An arched structure.

SOURCES OF ILLUSTRATIONS

STONE

p.12, E. Jervoise, the National Buildings Record, London; p. 13 (above) Emilia Bologna, courtesy the Harvard Architectural Library, Cambridge, Mass.; (below) *Kunst im Deutschen Reich*, Vol. 3, No. 8, Aug., 1939; p.14, Photo-Malina, Black Star, N.Y.; p.15, Folger Shakespeare Library, Washington, D.C.; p.16, E. Jervoise, the National Buildings Record, London; p. 17, Philip D. Gendreau, N.Y.; p.18, Alinari, Florence; p.19, Gauthey, Emiland Marie: *Oeuvres: Traité de la Construction des Ponts*, Navier, ed., Paris, Didot, 1809, Vol. 1, fig. 3; pp.20 and 21 (above) Palladio, Andrea: *The Architecture of A. Palladio*, London, Ward, 1742; p.21 (below) National Buildings Record, London; p.22, Philip D. Gendreau, N.Y.; p.23, E. Meerkämper, courtesy Swiss Federal Railroads, N.Y.; p.24 (above) Gauthey, Emiland Marie: *Oeuvres: Traité de la Construction des Ponts*, Navier, ed., Paris, Didot, 1809, Vol. I, pl.64; (below) Duplication Service, Library of Congress, Washington, D.C.; p.25 (above) *Theory, Practise and Architecture of Bridges*, John Weale, ed., London, The Architectural Library, 1843, Vol. IV, pl.58; (below) Philip D. Gendreau, N.Y.; p.26, courtesy Dr. Ssu-ch'eng Liang; p.27, E. Jervoise, the National Buildings Record, London; p.28, Berenice Abbott, N.Y.; p.29, *Kunst im Deutschen Reich*, Vol. 3, No. 8, Aug., 1939.

WOOD

p.30, American Museum of Natural History, N.Y.; p.31 (above left) courtesy Tennessee Valley Authority, Knoxville, Tenn.; (above right) The Bettmann Archive, N.Y.; (center right) Pope, Thomas: *Treatise on Bridge Architecture*, N.Y., printed for the author by A. Niven, 1811, pl.9; (below) Museum of Modern Art, N.Y.; p.32, Chinese News Service Photos, from Paul Guillumette, Inc., N.Y.; p.33, American Museum of Natural History, N.Y.; p.34 (above and below) courtesy *Engineering News Record*, N.Y.; (left) Fletcher, Robert and Snow, J.P.; *History of the Development of Wooden Bridges*, Paper #1864, ASCE *Transactions*, N.Y., 1934; p.35, Edmund H. Royce, from Congdon, Herbert Wheaton: *The Covered Bridge*, N.Y., Knopf, 1946; p.36 (above and below) Pope, Thomas: *Treatise on Bridge Architecture*, N.Y., printed for the author by A. Niven, 1811; p.37, courtesy C.L.V. Meeks, print in the William Barclay Parsons Collection at Columbia University, made from a drawing by G. A. Busby and published in London by Taylor in 1823; p.38, P. A. Dearborn, N.Y.; p.39, Fachklasse für Fotografie, Gewerbeschule Zürich, courtesy *Das Werk*.

METAL ARCH

p.40 (above) Gauthey, Emiland Marie: *Oeuvres: Traité de la Construction des Ponts*, Navier, ed., Paris, Didot, 1813, Vol. II, pl.V-1; (center and below) courtesy Henry-Russell Hitchcock, Jr., prints in the Science Museum, London, made from drawings by J. Raffield and published in London by Taylor in 1798; p.41, courtesy Henry-Russell Hitchcock, Jr., from an aquatint of 1801 in the Science Museum, London; p.42 (above) Telford, Thomas: *Life of Thomas Telford*, London, Payne and Foss, 1838; (center) E. Jervoise, the National Buildings Record, London; (below) courtesy American Bridge Company, Pittsburgh; p.43, ND, courtesy *L'Architecture d'Aujourd'hui*, Paris; p.44 (above) courtesy Hanover and Hardesty, N.Y.; (center) Underwood and Underwood, courtesy American Institute of Steel Construction, N.Y.; (below) *Die Schweizerische Bauzeitung*, 1913, courtesy Paul Zuberbühler; p.45, Rodney McKay Morgan, N.Y.; p.46 (below) courtesy O. H. Ammann; p.47, G. E. Kidder Smith, N.Y.; p.48, Hoyt, courtesy Port of New York Authority; p.49, courtesy Port of New York Authority; p.50 (above) courtesy American Institute of Steel Construction, N.Y.; (center and below) and p.51, courtesy State of New Hampshire Highway Department; p.52, courtesy American Institute of Steel Construction, N.Y.; p.53 (above) courtesy Chicago Park District; (below) Hedrich-Blessing Studio, Chicago.

SUSPENSION CABLE

p.55 (above) Navier, Claude: *Rapport et Memoire sur les Ponts Suspendus*, Paris, Imprimerie royale, 1823, pl. I; (below) courtesy Rudolf Mock; p.56 (above) Brunel, Isambard: *Life of I. K. Brunel*, London, Longmans Green, 1870, frontis.; (center and below) *The Architectural Review*, London, Sept., 1939, courtesy The Central Library, Bristol; p.57 (above) courtesy Henry-Russell Hitchcock, Jr.; (below) R. Wills, the National Buildings Record, London; p.58, courtesy New York Department of Public Works; p.59 (above) Keystone View Co., Inc., N.Y.; (below) courtesy Port of New York Authority; p.60 (above) courtesy Redwood Empire Association, San Francisco; (below) Ciné-photographie, Quebec, courtesy Quebec Department of Public Works; p.61, Rodney McKay Morgan, N.Y.; p.62, *Kunst im Deutschen Reich*, Vol. 6, No. 12, Dec., 1942; p.63, Wide World Photos, courtesy *Engineering News Record*, N.Y.

METAL BEAM

p.64, courtesy *Engineering News Record*, N.Y.; p.65, Schaechterle, Karl W. and Leonhardt, Fritz: *Die Gestaltung der Brücken*, Berlin, Volk und Reich Verlag, 1937; p.66, *Kunst im Deutschen Reich*, Vol. 3, No. 8, Aug., 1939; p.67 (above and center) courtesy Tennessee Valley Authority, Knoxville, Tenn.; (below) courtesy Triborough Bridge Authority, N.Y.; p.68 (above and center) courtesy American Bridge Company, Pittsburgh; (below) and p.69, courtesy Tennessee Valley Authority, Knoxville, Tenn.; p.70 (above) courtesy American Institute of Steel Construction, N.Y.; (below) courtesy Tennessee Valley Authority, Knoxville, Tenn.; p.71, Bureau of Reclamation, courtesy *Engineering News Record*, N.Y.; p.72 (above) Schaechterle, Karl W. and Leonhardt, Fritz: *Die Gestaltung der Brücken*, Berlin, Volk und Reich Verlag, 1937; (below) courtesy Port of New York Authority; p.73, Foto Gross, St. Gallen O., Switzerland, courtesy Max Bill, Zurich; p.74, British Information Services, N.Y., courtesy *Engineering News Record*, N.Y.; pp.75 and 76 (above and below) courtesy *Engineering News Record*, N.Y.; p.76 (center) International Commercial Photo Co., N.Y., courtesy American Institute of Steel Construction, N.Y.; p.77, courtesy *Engineering News Record*, N.Y.; p.78 (above) courtesy *L'Architecture d'Aujourd'hui*, Paris; (center) courtesy *Engineering News Record*, N.Y.; (below) courtesy American Institute of Steel Construction, N.Y.; pp.79, 80 (above) and 81, courtesy *Engineering News Record*, N.Y.; p.82, Ewing Galloway, N.Y.; p.83 (above) Möhringer, Karl: *The Bridges of the Rhine*, Baden, Germany, Joh. Möhringer Verlag, 1931, courtesy *Engineering News Record*, N.Y.; (center) courtesy *Engineering News Record*, N.Y.; (below) courtesy *L'Architecture d'Aujourd'hui*, Paris.

REINFORCED CONCRETE

p.84 (above) courtesy San Francisco Board of Park Commissioners; (below) courtesy *Engineering News Record*, N.Y.; p.85 (above) courtesy National Park Service, Washington, D.C.; (center) courtesy Connecticut Highway Department; (below) courtesy *Engineering News Record*, N.Y.; p.86, *Le Beton Armé*, Mar. 3, 1919, courtesy *L'Architecture d'Aujourd'hui*; p.87, Bureau Maillart, L. Meisser, Ing., courtesy Dr. Sigfried Giedion, Zurich.

REINFORCED CONCRETE ARCH

p.88 (above) courtesy Cuyahoga County Engineer, Ohio; (below) and p.89, courtesy California Department of Public Works; p.90 (above) Skånska Cement Company, Malmö, Sweden, courtesy G. E. Kidder Smith, N.Y.; (below) G. E. Kidder Smith, N.Y.; pp.91 and 92 (above) G. E. Kidder Smith, N.Y.; (below) courtesy R. E. Enthoven, Librarian, Royal Institute of British Architects; p.93, G. E. Kidder Smith, N.Y.; pp.94 and 95, courtesy Paul Zuberbühler; p.96, H. Baranger, Paris, courtesy *L'Architecture d'Aujourd'hui*, Paris; pp.97 and 98, Skånska Cement Company, Malmö, Sweden, courtesy G. E. Kidder Smith, N.Y.; p.99 (above) courtesy *L'Architecture d'Aujourd'hui*, Paris; (below) *The Architectural Forum*, N.Y.; pp.100 and 101, H. Baranger, Paris; p.102, courtesy Max Bill, Zurich; p. 103, O. Engler, Winterthur, Switzerland, courtesy Dr. Sigfried Giedion, Zurich; pp.104 and 105, Max Bill, Zurich; pp.106 and 107, Mischol, Schiers, courtesy Dr. Sigfried Giedion, Zurich; p.108 (above) Max Bill, Zurich; (below) H. Wolf-Bender's Erben, Zurich; p.109, courtesy Dr. Sigfried Giedion, Zurich, p.110 (above) P. Boissonas, Geneva; (center) H. Wolf-Bender's Erben, Zurich; (below) Max Bill, Zurich; p.111, P. Boissonas, Geneva.

REINFORCED CONCRETE: BEAM AND RIGID FRAME

p.112, Max Bill, Zurich; p.113, Ryner, courtesy Prader & Cie, Zurich; pp.114 and 115, Sunami, N.Y.; p.116 (above) courtesy Tennessee Valley Authority, Knoxville, Tenn.; (upper center) courtesy G. E. Kidder Smith, N.Y.; (lower center) courtesy *Engineering News Record*, N.Y.; (below) courtesy E. S. Elcock, p.117, courtesy *Engineering News Record*, N.Y.; p.118 (above) F. Zurcher, Lausanne, courtesy Paul Zuberbühler; (below) "Topical" Press Agency, London, courtesy *The Architectural Review*, London; p.119, Dell and Wainwright, courtesy *The Architectural Review*, London; pp.120 and 121, Sunami, N.Y.

THIS BOOK WAS PRINTED IN 1949

FOR THE TRUSTEES OF THE MUSEUM OF MODERN ART

BY MODERN GRAVURE CORPORATION, NEW YORK

COVER AND TYPOGRAPHY BY EDWARD L. MILLS

Museum of Modern Art Publications in Reprint

Max Ernst. 1961. William S. Lieberman

Fantastic Art, Dada, Surrealism. 1947. Barr; Hugnet

Feininger-Hartley. 1944. Schardt, Barr, and Wheeler

The Film Index: A Bibliography (Vol. 1, The Film as Art). 1941.

Five American Sculptors: Alexander Calder; The Sculpture of John B. Flannagan; Gaston Lachaise; The Sculpture of Elie Nadelman; The Sculpture of Jacques Lipchitz. 1935-1954. Sweeney; Miller, Zigrosser; Kirstein; Hope

Five European Sculptors: Naum Gabo—Antoine Pevsner; Wilhelm Lehmbruck— Aristide Maillol; Henry Moore. 1930-1948. Read, Olson, Chanin; Abbott; Sweeney

Four American Painters: George Caleb Bingham; Winslow Homer, Albert P. Ryder, Thomas Eakins. 1930-1935. Rogers, Musick, Pope; Mather, Burroughs, Goodrich

German Art of the Twentieth Century. 1957. Haftmann, Hentzen and Lieberman; Ritchie

Vincent van Gogh: A Monograph; A Bibliography. 1935, 1942. Barr; Brooks

Arshile Gorky. 1962. William C. Seitz

Hans Hofmann. 1963. William C. Seitz

Indian Art of the United States. 1941. Douglas and d'Harnoncourt

Introductions to Modern Design: What is Modern Design?; What is Modern Interior Design? 1950-1953. Edgar Kaufmann, Jr.

Paul Klee: Three Exhibitions: 1930; 1941; 1949. 1945-1949. Barr; J. Feininger, L. Feininger, Sweeney, Miller; Soby

Latin American Architecture Since 1945. 1955. Henry-Russell Hitchcock

Lautrec-Redon. 1931. Jere Abbott

Machine Art. 1934. Philip Johnson

John Marin. 1936. McBride, Hartley and Benson

Masters of Popular Painting. 1938. Cahill, Gauthier, Miller, Cassou, et al.

Matisse: His Art and His Public. 1951. Alfred H. Barr, Jr.

Joan Miro. 1941. James Johnson Sweeney

Modern Architecture in England. 1937. Hitchcock and Bauer

Modern Architecture: International Exhibition. 1932. Hitchcock, Johnson, Mumford; Barr

Modern German Painting and Sculpture. 1931. Alfred H. Barr, Jr.

Modigliani: Paintings, Drawings, Sculpture. 1951. James Thrall Soby

Claude Monet: Seasons and Moments. 1960. William C. Seitz

Edvard Munch; A Selection of His Prints From American Collections. 1957. William S. Lieberman

The New American Painting; As Shown in Eight European Countries, 1958-1959. 1959. Alfred H. Barr, Jr.

New Horizons in American Art. 1936. Holger Cahill

New Images of Man. 1959. Selz; Tillich

Organic Design in Home Furnishings. 1941. Eliot F. Noyes

Picasso: Fifty Years of His Art. 1946. Alfred H. Barr, Jr.

Prehistoric Rock Pictures in Europe and Africa. 1937. Frobenius and Fox

Diego Rivera. 1931. Frances Flynn Paine

Romantic Painting in America. 1943. Soby and Miller
Medardo Rosso. 1963. Margaret Scolari Barr
Mark Rothko. 1961. Peter Selz
Georges Roualt: Paintings and Prints. 1947. James Thrall Soby
Henri Rousseau. 1946. Daniel Catton Rich
Sculpture of the Twentieth Century. 1952. Andrew Carnduff Ritchie
Soutine. 1950. Monroe Wheeler
Yves Tanguy. 1955. James Thrall Soby
Tchelitchew: Paintings, Drawings. 1942. James Thrall Soby
Textiles and Ornaments of India. 1956. Jayakar and Irwin; Wheeler
**Three American Modernist Painters: Max Weber; Maurice Sterne; Stuart
    Davis.** 1930-1945. Barr; Kallen; Sweeney
**Three American Romantic Painters: Charles Burchfield: Early Watercolors;
    Florine Stettheimer; Franklin C. Watkins.** 1930-1950. Barr; McBride; Ritchie
**Three Painters of America: Charles Demuth; Charles Sheeler; Edward
    Hopper.** 1933-1950. Ritchie; Williams; Barr and Burchfield
Twentieth-Century Italian Art. 1949. Soby and Barr
Twenty Centuries of Mexican Art. 1940
Edouard Vuillard. 1954. Andrew Carnduff Ritchie

The Bulletin of the Museum of Modern Art, 1933-1963. (7 vols.)